WITH PREFACE BY DR CYNTHIA
Head of the Climate Impacts Group at the Columbia Center for Climate Systems Research

WAKING UP
TO
CLIMATE
CHANGE

Five Dimensions of the Crisis and What We Can Do About It

GEORGE ROPES

ClimateYou, USA

World Scientific

NEW JERSEY · LONDON · SINGAPORE · BEIJING · SHANGHAI · HONG KONG · TAIPEI · CHENNAI · TOKYO

Published by

World Scientific Publishing Co. Pte. Ltd.
5 Toh Tuck Link, Singapore 596224
USA office: 27 Warren Street, Suite 401-402, Hackensack, NJ 07601
UK office: 57 Shelton Street, Covent Garden, London WC2H 9HE

Library of Congress Cataloging-in-Publication Data
Names: Ropes, George, author.
Title: Waking up to climate change : five dimensions of the crisis and what we can do about it /
    George Ropes, ClimateYou, USA ; with preface by Dr. Cynthia Rosenzweig,
    Head of the Climate Impacts Group at the Columbia Center for Climate System Research.
Description: New Jersey : World Scientific, [2022]
Identifiers: LCCN 2022003633 | ISBN 9789811246234 (hardcover) |
    ISBN 9789811247545 (paperback) | ISBN 9789811246241 (ebook for institutions) |
    ISBN 9789811246258 (ebook for individuals)
Subjects: LCSH: Climatic changes.
Classification: LCC QC903 .R64 2022 | DDC 363.738/74--dc23/eng20220321
LC record available at https://lccn.loc.gov/2022003633

British Library Cataloguing-in-Publication Data
A catalogue record for this book is available from the British Library.

All photographs copyright by Andrew Dillon Bustin, used with permission.

For any available supplementary material, please visit
https://www.worldscientific.com/worldscibooks/10.1142/12525#t=suppl

Desk Editor: Amanda Yun

Typeset by Stallion Press
Email: enquiries@stallionpress.com

WAKING UP
TO
CLIMATE
CHANGE

Five Dimensions of the Crisis and What We Can Do About It

Dedication

This book is dedicated to my father, Dr. George H. Ropes, a math educator who lived in Westchester County, New York. He loved showing students alternative ways to approach and master hard concepts like long division and decimals. He was an early adopter of computers, purchasing a RadioShack TRS-80 computer soon after its launch in August 1977. He hitched it up to a modem so he could communicate with others, even though back then transmissions were painfully slow.

As the technology progressed, Hardy, as he was known, kept up with new advances. Early on, he advocated that the schools where he taught buy computers, if not for every student, at least enough to set up a "Math Laboratory." It was a hard sell in those days, but he persevered. Today, computers are everywhere — in schools, homes, businesses, and, of course, pockets. Interconnectivity is easy, fast, and a given. Hardy would be amazed and delighted that almost everyone can now be in touch with almost everyone else, sharing knowledge and ideas.

It was in his capacity as an educator that my father created one of the first interactive software programs in 1990 to teach students about climate change. Called *Hothouse Planet*,[1] it enabled students to understand the greenhouse effect and its implications for the future of our planet. Students conducted virtual experiments to test how world population, fossil fuels, and trace gases in the atmosphere affect temperatures and sea levels based on NASA data and projections. The program helped students and teachers

---

[1] Hothouse Planet, 1990, EME Corporation.

develop awareness of the importance of actions to minimize the growth rates of greenhouse gases, and it allowed users of the program to observe the effects of future temperature and sea level rise both locally and globally.

My father would applaud the efforts of the ClimateYou Alliance to link people of all ages from around the world to explore, learn about, and communicate the many ways that the changing climate impacts life today — and threatens the future of life as we know it, if we do not continue to wake up and act together more concertedly soon.

Preface

This volume deserves a prominent place in the burgeoning "bookshelf" on climate change. It tracks the build-up of attention to the most significant planetary issue of our time and delineates its five key dimensions. It presents the timeline of key events in the trajectory of climate change knowledge and action, as well as by including a series of commentaries on articles from public media linked to the dimensions. The articles span the timeframe from 2015 to 2022, the period in which climate change has exploded to nearly universal consciousness.

The commentaries in this book are all written in the powerful voice of George Ropes, senior editor of ClimateYou.org, one of the earliest and most longstanding social media platforms dedicated to the issue. From its founding over a decade ago by my brother George Ropes and myself, ClimateYou. org has been informing the general public about climate change. It has reached countless people and worked with hundreds of students through the educational programs of its parent organization, the ClimateYou Alliance.

The book defines and explores the five key dimensions of climate change: energy; weather and climate; consequences for nature and people; laws and leaders; and finance.

Energy is at the core of the climate crisis, as our civilization has been powered since the Industrial Revolution by fossil fuels — coal, oil, and natural gas. Changes in climate and weather are the atmospheric consequences of the greenhouse gases produced by the burning of these fuels. Impacts are the subsequent consequences wreaked by the changes in climate. Turning to solutions, the book then describes the history of the policies, politics, finances, and economics that have hindered responses and that are now helping to accelerate the necessary changes.

As a scientist who has worked on climate change impacts over the last several decades, I find it gratifying that the people throughout the world are now indeed "waking up" to the climate crisis. This book explains clearly how and why that is occurring and documents the ongoing trajectory of the issue.

I recommend this book to you wherever you are on your own path in becoming aware of and responding to climate change.

Dr. Cynthia Rosenzweig
Head, Climate Impacts Group
Columbia Center for
Climate Systems Research

Contents

Introduction

Waking Up to Climate Change

Skyscape (Costa Rica, 2009)

The Climate Alarm is Ringing

Climate science and public awareness have come a long way since early researchers in the 1880s discovered the heat-trapping role of carbon dioxide (CO_2) in the atmosphere. They correctly postulated the future climate effects of the coal-powered Industrial Revolution. This book is about waking up to the climate crisis, which is now upon us many decades later (see Table 1). To show the course of this awakening and look ahead, the book draws on writings from the ClimateYou.org website from 2015 to 2022.[1]

After World War II, Americans fully embraced the automobile, and the nation's first national highway system was built. In the economic boom that followed, the United States speedily industrialized, converting wartime factories to domestic production. The energy that fueled this expansion was carbon-rich organic plant material, buried for millions of years until it turned into coal, oil, and natural gas.

When burned, these fossil fuels produce emissions of carbon dioxide (CO_2), methane (CH_4), and nitrous oxide (N_2O), that together are called greenhouse gases (GHGs). In much the same way a greenhouse does, these gases trap heat in the atmosphere that would otherwise be radiated back into space. Spew enough GHGs into the atmosphere, as we have with our cars, trucks, airplanes, electricity generation, fracking, factories, farms, and home heating and cooling systems, and the Earth will warm up.

In the 1950s, scientists measuring atmospheric CO_2 on Mauna Loa Observatory in Hawaii were startled by its rapid increase. In 1988, Jim Hansen testified on the dangerous links between increasing carbon dioxide and global warming to the U.S. Congress, ringing the tocsin to alert the senators of an impending crisis. Although Hansen's testimony got substantial media coverage, America continued on its profligate way.

For a long time, people either ignored the effects we were having on the climate or they confused those effects with the weather, which varies daily. In recent years, however, more and more evidence of a changing climate has intruded itself upon our lives. Years-long drought and devastating fires have hit California. Repeated torrential rains in the U.S. Midwest have delayed planting the corn and soy crops past any expectation of breaking even or have destroyed crops already planted.

---

[1] See footnotes throughout the book and Appendix 1 for full list of sources.

Table 1.   Waking Up to Climate Change

1880s	Early scientists discover heat-trapping role of CO_2 in the atmosphere.
1950s	Scientists measure rapid increase in CO_2 at Mauna Loa, Hawaii.
1970s	United Nations holds its first climate conferences.
	Oil and coal companies write internal memos regarding CO_2 threat.
1980s	UN forms Intergovernmental Panel on Climate Change (IPCC).
	James Hansen gives testimony to the U.S. Senate that climate change is upon us.
1990s	United Nations holds first Conference on Environment and Development, also known as the Rio de Janeiro Earth Summit.
	United Nations establishes Framework Convention on Climate Change.
2000s	Al Gore presents the film *An Inconvenient Truth*, raising international public awareness of global warming.
2010s	Governments adopt Paris Agreement to limit warming to well below 2°C (3.6°F).
	Extinction Revolution is founded in the UK.
	Greta Thunberg leads School Strike for the Climate.
	Scientists declare climate crisis.
2020s	Extreme climate events multiply throughout the world.
	We begin a decade of action?

In the fall of 2012, Superstorm Sandy devastated Atlantic coastal areas from the Bahamas to Canada, causing catastrophic damage to the New York and New Jersey coasts. In 2017, Hurricane Harvey inundated Texas and Louisiana, Hurricane Irma left Florida and the Virgin Islands reeling, and Hurricane Maria wreaked havoc on Puerto Rico. In 2018, Hurricane Florence caused severe damage in the Bahamas and flooded the Carolinas, and Hurricane Michael flattened eastern Florida. And 2022 brought Hurricanes Fiona and Ian. Scientists are finding that such severe tropical storms have been increasing in recent decades. In each case, these storms left behind significant destruction (due in part to increasing development as well as to the changing climate), with effects lasting for months and even years.

Recent polls show that more than 70% of Americans believe climate change is real; more than 50% believe that we humans are the principal cause of the changing climate. It is not just the weather that has convinced so many Americans to believe that their climate is changing. Climate scientists, also called climatologists, are academics and are often loath to provoke controversy. For a long time, they tended to shun publicity rather than seek it. However, over the last decade, interest in and acceptance of climate science has grown dramatically, spurred by writers for such magazines as the *New Yorker*, *Atlantic*, *Mother Jones*, and *Rolling Stone* and by some scientists themselves. Several websites began to print the occasional article about climate change. More people were hearing about the climate, and some were becoming concerned about it. I was one of them.

My sister Cynthia Rosenzweig, a climate impacts scientist, and I began ClimateYou.org, a climate-related blog, in 2008. We conceived of the blog as a way to disseminate information to people interested in learning about climate change and those already somewhat knowledgeable who wanted to learn more. Back then, very few articles about climate change appeared on the web. Many days, I could not find any. Over the past decade, the coverage of climate change has increased exponentially. These days, there are often too many climate-related articles on any given day to include or reference them all on ClimateYou.org.

In the early years, my practice was to abstract each article, detailing the authors' research, their arguments, and their conclusions. Sometimes, I commented on the significance of their findings. In the blog's middle years, as the number of climate studies and reports began to grow, I wrote fewer full abstracts. Instead, I adopted a shorter form in which I tried to capture the essence of each piece, in addition to providing the relevant link. The aim was to present, enough information to let readers decide whether to read the entire article for all the details of the new finding or whether the summary would suffice.

In 2010, the ClimateYou.org blog expanded its staff and internet presence, bringing on board the accomplished environmental journalist, Abby Luby, as Senior Editor. At her suggestion, we added a thread, "Our Take," to accommodate longer, more personal opinion pieces and to field more responses from readers. At the same time, Abby helped to broaden our outreach on Facebook and Twitter, posting items and comments to each.

Given Twitter's strict character limit, my comments there had to be pithy. Most items destined for posting on social media have one line of comment, but some have more, three to six lines, providing a short summary of the item and its significance. Items with longer comments get funneled to Facebook.

This book is a collection of my ClimateYou.org writings through the years that together chart the process by which we have woken up to the most significant issue of our time: climate change. In many cases, they have been updated with the latest information about the topics they cover. The goal of bringing these up-to-date pieces into a book is to help people become aware of the many dimensions of climate change, understand how the climate change issue is evolving into action, and become pathfinders to solutions.

*****

The climate is complex. More than a simple system, it is a system of systems. Each system influences others, and in turn is influenced by them. It is often hard to know where one system ends and another begins. With the help of the ClimateYou.org team, I faced this problem when trying to organize hundreds of articles into key dimensions and to select the most important and relevant ones. Both tasks gave me a new appreciation of the profound and multifaceted effects of climate change. It affects every dimension of life.

Climate change is, at its core, an *energy* story. **Dimension 1: Energy** looks at how the burning of fossil fuels to generate power has created the mess we are in. A swift transition from fossil fuels to renewables is clearly critical to avoiding the most catastrophic consequences of climate change. The first section on "Coal, Oil, and Gas" describes the status and trends of the continued use of fossil fuels. About two-thirds of natural gas in the U.S. is obtained by fracturing or "fracking" the shale in which it is embedded. Some of the gas is sent through pipelines to energy plants, and some is made into liquefied natural gas (LNG) for long-distance transport. This section asks and answers a key climate change question — Is there an end to using coal?

The section on "Energy Sources in Transition" considers developments in renewable sources (wind, solar, hydro, geothermal, and biogas) and the state of nuclear energy today. Other aspects of energy considered in Dimension 1 are "Transportation" (with gasoline-powered vehicles at present one of the largest contributors to climate change) and "Innovation and Technology." There is tempered optimism that technological advancements can contribute both to developing new, better energy sources and to speeding the needed transition to their widespread adoption.

Dimension 2: Weather and Climate considers the confusion that has existed between these two terms. Weather is variable day to day — hot today, cold tomorrow, sunny the day after, rainy all weekend. Climate, on the other hand, is long term, its trends only visible over 30 years or more. The section on "Extreme Weather" examines how some weather phenomena that have always been around — droughts, floods, storms, heat waves, and wildfires — have been supercharged by climate change, creating new and evermore dangerous threats. This dimension also looks at "Oceans and Coasts," which are especially affected by climate change. Global heating is expanding the water in the oceans and is melting the icesheets at both poles. Together, these phenomena are causing the sea level to rise, leading to exacerbated flooding of coastal towns and cities.

Oceans cover 70% of the planet, and they have profound effects on the climate. Their health is our health. The Atlantic Gulf Stream carries warm water from the tropics to northeast North America and northern Europe. The west coast of South America occasionally experiences warm currents known as El Niño events, which are due to intertwined atmospheric and oceanic processes over the Pacific. El Niño tends to bring warm, wet weather to the west coasts of the Americas, and dry weather to Australia and Asia. La Niña does the reverse. How climate change is affecting these ocean-related processes is an active and important area of research.

Climate change is pervasive. It impacts every aspect of our lives. **Dimension 3: Consequences for Nature and People** covers several of the most important areas of change: "Deforestation, Fires, and Species Extinction;" "Famine and Food Security;" "Cities and Power Outages;" and "Migration, Conflict, and Population." Looking at the effects on these major natural and social components, it becomes clear that no iota will be left unaffected by climate change. This third dimension is about the large-scale impacts that are coming, that have already come, and that will intensify as the Earth continues to warm.

Dimension 4: Laws and Leaders addresses the challenges faced by governments and international bodies and the people who head them as they work separately and together to solve the dilemma of climate change. Since climate change is a global phenomenon, it affects the citizens of every country. The first section, "International Action," describes the efforts to tackle these challenges through policy. The pathway to international agreement on climate change was excruciatingly slow but finally culminated in the Paris Agreement in 2015 (see Appendix 2, Table 1). Section 2 on "Ups and Downs: The Role of the United States" shows that U.S. leadership on climate change has made both advances and retreats (see Appendix 2, Table 2). The section on "China, India, Europe, and the Oil Producers" describes the roles of these crucial actors in developing and implementing the policies that are needed to solve climate change and the challenges that these countries face. The final section of Dimension 4 highlights some of the "Heroes" who have guided the world to wake up to climate change.

Most governments have committed to achieving the targets of the Paris Agreement, but they are lagging in taking the actions required to reach them. Only a few nations have put a price on carbon, which would make it more expensive to emit greenhouse gases, thereby reducing emissions. Few, if any, nations have ended the massive subsidies accorded to national fossil-fuel industries.

Dimension 5: Finance explores the realm of corporate actions on climate. Private sector responses are just as essential as government policies. Companies, regardless of what sector they are in — extractive, industrial, or service — need to review their operations with a view toward increasing energy efficiency and lowering greenhouse gas emissions. Many companies, now over 400 in the U.S., have committed to target dates for reaching net-zero emissions. However, the fossil fuel industry continues to be a juggernaut. It is sustained by government subsidies of many billions of dollars. Exxon alone has a market capitalization of $265 billion.

Recently, hints that the industry may be forced to change have appeared. Not only has public opinion turned against Big Oil, but the banks, sovereign funds, pension funds, international lenders like the World Bank and the International Monetary Fund, as well as the large insurance companies, are beginning to rein in the fossil fuel industry, restricting their loans and imposing performance-based conditions. The fossil fuel industry is like one of their own supertankers — hard to turn around. Still, the effort has begun and will accelerate. The politicians may drag their feet and shirk their responsibilities to the people and to the environment, but financiers have a keen eye for risks to their rewards.

The concluding section of the book is **We Are Waking Up to Climate Change — Now What?** It weaves together all five dimensions and focuses on solutions and ways to take action. Climate change is now recognized as a climate crisis. It is time for all actors — individuals, universities and research labs, corporations, governments, and civil society — to examine their practices, assess the risks of inaction, and chart their way forward to contribute to climate change solutions. These contributions need to act on two fronts: mitigation of greenhouse gas concentrations in the atmosphere and enhancement of resilience to climate change impacts. (See Appendix 3 for organizations working on climate change around the world.)

What can individuals do to ameliorate climate change effects? Lists of ways to reduce one's carbon footprint abound. On a personal level, such actions signal a commitment that helps to develop a community sense that everyone is doing their part to solve the crisis. Turning out the lights when we leave a room lowers our energy usage so the carbon emissions we cause go down. Recycling paper and some plastic products helps, too. Eating less meat, or none at all, can help to change the economics and practices of the livestock industry. Walking or bicycling, rather than driving, to do errands locally cuts our carbon footprint. Flying less often or not flying at all can reduce it even more.

Put simply and collectively, the four main tasks for all people to undertake together are:

1. Curtail CO_2 and other greenhouse gas emissions from fossil fuels;
2. Restore organic matter in sorely depleted soils;
3. Reforest cleared lands to build up carbon stores; and
4. Develop resilience to climate change now and in the future.

Many have started to wake up to climate change and are taking action. But the scale of the challenge is great, and there is still much more to do.

Addendum: Much of this book was prepared during the coronavirus pandemic of 2020–2022. COVID-19, an acute systemic threat, took over the world, bringing death to millions, while climate change, a long-term systemic threat, continued on its seemingly inexorable path. Even while the pandemic is still raging in many places throughout the world as I write this, I believe that we must build climate change solutions into the world's recovery from the coronavirus pandemic. We must bounce forward to create a truly climate-friendly world, not bring back the energy systems of the past that led to the climate crisis.

Pandemic concerns rapidly replaced climate-related ones, much to the consternation of activists, who worried that the momentum that had been building for climate action might dissipate irretrievably. Quarantined and isolated, we were heartened, however, by an unexpected result of the global economic shutdown. The skies cleared over city after city — Beijing, Los Angeles, London, Delhi, and hundreds more. The novelty of blue skies and lower CO_2 emissions made millions of people aware of the pollution caused by our dependence on fossil fuels. This awareness led to many voices advocating for a green recovery, not a return to a fossil fuel-based economy.

Now in 2022, war has broken out in Ukraine. Russia's invasion of its neighbor throws a harsh spotlight on how threats to world stability and security can suddenly arise and lead to other immense perils. For example, the conflict in Ukraine is disrupting energy and food supplies in many parts of the globe.

We must now learn to confront simultaneously the triple threats of climate change, pandemic disease, and brutal conflict. The world is finally awake to the climate crisis enveloping us all, even as the scourge of the coronavirus continues to spread and re-spread, and war is dislocating millions of people in Europe. Many individuals and groups are clamoring with an urgency that must be heeded for greater, more effective, and faster action on all fronts. The mandate is to preserve the well-being of humanity and the Earth, our only home, forever.

Dimension 1

Energy

City lights from speeding cab (New York, 2010)

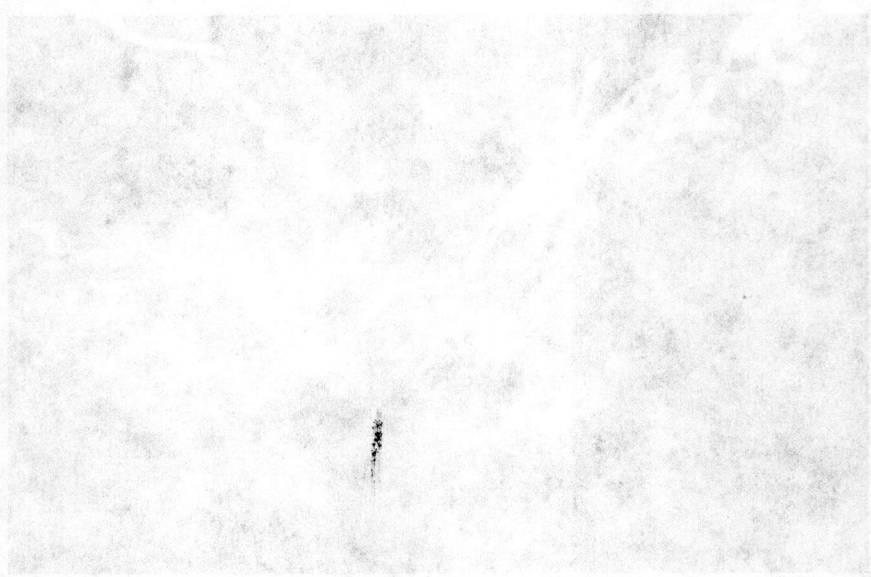

Energy

Climate change is very much an energy story. Energy from the burning of fossil fuels — coal, oil, and gas — has caused the crisis we are in now. An accelerated transition from fossil fuels to renewables is essential to reduce the harm that climate change otherwise will bring. That transition is central to halting the emissions of greenhouse gases that are responsible for the changes to our climate that threaten to make life on Earth difficult, if not impossible.

As many climate scientists have proven with voluminous, uncontestable data, the situation is dire and growing ever more so. On the plus side, developments in renewable energy sources — wind, solar, hydro, geothermal, and biogas — abound and are gradually being more widely adopted. On the negative side, the main fossil fuels — coal, oil, and natural gas — are by far still the main sources of energy investments.

Section 1 of the Energy Dimension tracks what is happening with "Coal, Oil, and Gas." Gas obtained by the hydraulic fracturing of shale rock deep underground in a process known as fracking has become an increasingly important source of natural gas in the United States since the middle of the last century. Interest in exploiting gas shale formations has spread to the rest of the world. Because it is found trapped within fine-grained sedimentary rocks, millions of liters of water and hundreds of thousands of liters of chemicals are often injected into the subsurface as part of its extraction.

Section 2 covers "Energy Sources in Transition." Technological developments have enabled, even driven, the transition from fossil fuel energy to renewable energy from the wind and the sun, to the extent that transition has occurred. Wind turbines have seen blade and power-train design improvements that have enabled larger and more efficient operation. These turbines are now capable of delivering more energy at a cost equal to or less than any fossil fuel.

Solar panels have likewise plummeted in price to the point where they too are cost-competitive with fossil fuels. Current silicon-based photovoltaic cells (PVCs) are not much more efficient than previous ones, but the gains in lowered production costs of cell panels and in the construction costs of solar "farms," covering many acres, have powered solar energy's rapid deployment. Several labs are working to develop more efficient PVCs using alternate photosensitive materials, such as perovskites (materials with the same crystal structure as calcium titanium oxide), or ones that can be painted or sprayed on walls and roofs. Researchers working on double-sided PVCs and ones that work at night have notched recent advances.

Both wind and solar have one big shortcoming: neither produces energy continuously. The wind does not always blow, nor does the sun shine at night. The solution to both these intermittency problems is to store energy when it is abundant for when it is needed but not being generated. Batteries provide that capacity, and their technological development has only lagged behind that of wind and solar by a little. Today's batteries can store more energy for longer and at lower cost than those of just a few years ago; recent advances promise that tomorrow's batteries will be even better. Systems that combine wind or solar with storage may supplant nearly all fossil fuel energy in the foreseeable future.

Section 3 on "Transportation" explores this major source of greenhouse gas emissions, with vehicles producing roughly one-third of all the carbon emissions in the United States. Anticipating changes in policies, mandates, incentives, and market conditions, nearly every automaker offers at least one model of an electric vehicle (EV) or has unveiled new models with technological advances. Tesla has seized the EV lead with its Models S, X, 3, and Y. Whether it can retain its dominance depends in part on Elon Musk's business acumen and in part on the competition. Chevrolet makes the Bolt EV, a battery-electric car with about 250 miles of range, and has plans for other electric vehicles, minivans, and SUVs.

China's BYD, Hong Kong's Byton, Germany's Volkswagen, Audi, and BMW, and companies in Japan, the UK, France, Sweden, and South Korea are in the race. Beyond the purely economic factors involved in designing, testing, and producing EVs, all are affected by the need for more charging stations and more efficient, lower-cost, faster-charging batteries, and by consumers' buying decisions. As of 2021, 11% of new cars sold in the U.S. and about 9% of new cars sold in the world were electric vehicles. These percentages are expected to rise, but by how much or how soon is uncertain.

Section 4 covers "Innovation and Technology." Energy innovation is not limited to wind and solar. Scientists can now split water into its component parts hydrogen and oxygen cheaply and without generating any carbon emissions. This so-called "green hydrogen" is made by passing an electrical current made from renewable energy. The hydrogen goes into fuel cells that generate energy, with water the only by-product. Although hydrogen fuel cell (HFC) technology is in its infancy, HFC vehicles have already been designed and piloted. It will not be many years before electric vehicles may share the roads with hydrogen-fueled ones.

Transforming our energy system is crucial to solving climate change.

Coal, Oil, and Gas

Sunoco oil refinery (New Jersey, 2008)

Is There an End to Using Coal?

It was really quite depressing to learn in 2019 that the International Energy Agency (IEA) expected coal demand to be stable for the next four years.[1] Coal demand is falling in the U.S. and EU but is surging in Asia, especially in India and China, already two of the top three emitters of the heat-trapping gas CO_2. Both have large populations and fast-growing economies. Many of their coal plants are new, with long high-emission lifetimes ahead of them. The world cannot afford their future emissions. They will propel the world past the limit required to keep temperature rise well under the 2° Celsius (3.6° Fahrenheit) target set in the Paris Agreement that must be heeded to avoid the most catastrophic effects of climate heating.

Were politicians to act more responsibly, no more coal plants would be built, and existing ones would be strictly regulated to minimize their emissions. In the absence of political will, however, extraordinary, supranational measures may be required. One approach would be a global tax on carbon emissions, which would fall on all fossil fuels, but most heavily on coal, the "dirtiest" fossil fuel. Another idea that has been floated is a border tax, imposed by countries that have invested in clean power sources, on countries that have not done so but want to export their cheaper "dirty" goods to countries that have.

Both approaches are controversial, and both face high bars to becoming operational. However, the alternative is a hostile, even deadly environment for the people of nearly all the world's countries. That climate-changed world is so bad that it makes the politically difficult possible, no matter how unattractive. While a coal-fouled future as predicted by the IEA is depressing, there is at least a glimmer of hope that apocalypse can be avoided.

---

[1] https://www.iea.org/reports/coal-2019, Coal 2019: Analysis and Forecast to 2024, IEA.

Coal Is Dying

Coal is a dead man walking, but it is taking too long to die. As the dirtiest of fossil fuels, coal is one of the main reasons why the climate is in crisis. The CO_2 it emits is a major cause of global heating, and the carbon particulates from the combustion of coal contribute to the eight million deaths globally each year from air pollution.

Growing universal awareness that the climate is in crisis has increased pressure on the coal industry, the banks that finance it, and the insurance companies that insure it. All have begun reacting in order to limit their risk exposure. Coal is also threatened economically, as natural gas and both wind and solar now beat coal on price.

The industry should be dead, but new plants continue to be planned, financed, and approved. Plants built today will speed emissions and pollutants for 30 or 40 years. Yet, by that time, the world must be close to net-zero emissions to avoid the most catastrophic consequences of a climate gone hostile.

The coal industry knows it is ailing, but it does not yet accept that its decline is irreversible, that its case is terminal. The financial sector has set policies to curb it, such as loan eligibility rules and rates, but loopholes often render them ineffective. Governments are unwilling or unable to disrupt an established industry that employs thousands, generates the power that drives their economies, and supports many civil servants. Despite the best efforts of climate activists such as Greta Thunberg,[2] political will is sorely lacking. Even stern warnings from the Secretary General of the United Nations, Antonio Guterres, have had little effect.[3]

Market forces are working, finance is starting to squeeze, and public disapprobation is tarnishing the industry's image, yet none of these forces is acting quickly or forcefully enough. The climate crisis will not wait.

---

[2] https://www.youtube.com/watch?v=TMrtLsQbaok, "Greta Thunberg to world leaders: 'How dare you? You have stolen my dreams and my childhood,'" *Guardian News*, YouTube, September 23, 2019.

[3] https://www.un.org/press/en/2019/sgsm19584.doc.htm, "To Tackle Climate Change, Leaders Must Tax Pollution, Not People, End Coal Plant Construction by 2020, Secretary-General Urges, Concluding Pacific Region Visit," United Nations, May 18, 2019.

The next chance for a coal-free world was at the United Nations Framework Convention on Climate Change 26th Conference of the Parties November 2021 in Glasgow, Scotland. At this UN Climate Summit, six years after the Paris Agreement,[4] member nations gathered to commit to greater efforts to keep global warming to tolerable limits.

At COP26, an agreed statement called upon nations to "phase down" unabated coal power and inefficient subsidies for fossil fuels. While this statement was weaker than many countries were calling for, it was the first time that the coal industry was called out explicitly in a COP agreement. With the re-engagement of the U.S. under the Biden administration, the nearly 200 nations of the world got an infusion of new hope. Can this new hope counter humanity's innate propensity to discount possible future rewards in favor of either immediate gains or the entrenched power of the coal industry?

Fracking: Biggest Cause of Massive Methane Rise?

Evidence that fracking has led to more methane in the atmosphere is compelling. More methane means more global warming. An article in *National Geographic* titled "Fracking boom tied to methane spike in Earth's atmosphere," written by Stephen Leahy,[5] documents the massive increase in methane emissions detected when fracking occurs.

Tighter regulation on the venting of gas from fracking wells is clearly indicated. Outright banning of fracking should also be considered. Several states (New York, Maryland, and Vermont) and countries (France and Germany) have already done so.

However, there are signs that the fracking boom may be reaching an end. In the Permian Basin in West Texas and Southeast New Mexico, Tier 1

---

[4] https://unfccc.int/process-and-meetings/the-paris-agreement/the-paris-agreement, The Paris Agreement.

[5] https://www.nationalgeographic.com/environment/article/fracking-boom-tied-to-methane-spike-in-earths-atmosphere, "Fracking boom tied to methane spike in Earth's atmosphere," Stephen Leahy, *National Geographic*, August 16, 2019.

wells, i,e., the largest, most profitable ones, are becoming harder to find and exploit. Many fracking companies, already dependent on the price of oil for their profitability and on investors' confidence in their continued robust returns, may soon be filing for bankruptcy. When that happens, the world will breathe a little easier.

Methane impacts climate change mightily.

The Questionable Future of Shale Gas

Over the last decade, the combined technologies of horizontal drilling and hydraulic fracturing (fracking) have allowed access to large volumes of shale gas that were previously uneconomical to extract. These technologies have given the natural gas industry in the United States a tremendous boost. Daniel Yergin and Samuel Andrus are respected figures in the energy field, and they describe the dramatic impact of recent shale gas production on the U.S. energy industry in a report published by IHS Markit.[6] However, their prediction of continued rapid growth over the next 20 years is on shakier ground.

The global energy system is in the midst of a transition from energy based on fossil fuels, which all produce greenhouse gas emissions (GHGs), to energy based on renewable energy sources (mostly wind, solar, and hydro), which do not. GHGs trap heat from the sun, like a greenhouse does, and warm the planet, thereby setting off climate change. Wind and solar energy are making inroads into fossil fuels markets, often beating them on price. Of the fossil fuels, coal — the dirtiest energy source in the sense of producing the most GHGs — is already under siege as an industry and an energy source. However, the worldwide industry is putting up a major fight to keep coal in business.

Oil from tar sands is a mixture of sand, clay, water, and a hydrocarbon substance called bitumen, which is used to produce gasoline and other petroleum products. Oil sands are the next dirtiest energy source after

---

[6] https://ihsmarkit.com/research-analysis/The-Shale-Gales-10th-birthday-present-massive-economic-resource-estimate.html, "The Shale Gales 10th birthday present, a massive economic resource estimate," Samuel Andrus, *IHS Markit*, March 20, 2018.

coal. The industry that produces tar sand oil has only a few years of viability before its markets dry up, given its costs to mine and refine, and its high GHG content.

Oil from Saudi Arabia, Iran, Venezuela, the Permian Basin in Texas, the Gulf of Mexico, and elsewhere has less expensive production costs and lower GHG content than coal. Therefore, oil from these sources might have longer product viability. However, it too is ultimately doomed as an energy source, as worldwide efforts to decrease GHGs finally become more urgent and effective. The vast untapped oil reserves of nation states and the major oil-producing companies are destined to become worthless, uneconomical to produce.

Natural gas, the product of the shale-fracking revolution, is cheaper and somewhat cleaner-burning than other fossil fuels, but its production and transport emit significant quantities of methane, a greenhouse gas more powerful than CO_2. Whether natural gas has another 20 years to run is far from certain. Much depends on the course of climate change, how bad conditions get and how fast, and on the global reaction, i.e., the pressure the public puts on governments and corporations to rein in GHG emissions. The future of shale gas also depends on how fast financial institutions — banks, pension funds, and stock markets — invest in the fossil-fuel sector — and to what extent they divest from it.[7]

---

[7] https://www.forbes.com/sites/williampentland/2018/07/07/shale-gas-revolution-is-just-getting-started/?sh=229e6bc37c11, "Shale Gas Revolution Is Just Getting Started," William Pentland, *Forbes*, July 7, 2018.

Energy Sources in Transition

Solar panels on earthen home (Colorado, 2019)

Using More Renewable Natural Gas

In the United States, the industrial use of renewable natural gas (RNG) is growing, including in the shipping industry. RNG is made by capturing biogases from decomposing organic material from sewage, agricultural production, food preparation, and residential lots. Its total lifecycle greenhouse gas emissions are much lower than those of natural gas, a fossil fuel. The use of RNG produces radically less carbon dioxide, carbon monoxide, and nitrogen oxide emissions than traditional propane, gasoline, or diesel-fueled vehicles.

Because of RNG's expanding use, it is becoming more viable and cost efficient. Energy Vision, a sustainable energy organization based in New York City, has recognized several companies with awards for using RNGs.[8]

- *United Parcel Service*: In 2015, the delivery company entered into a purchase agreement with Clean Energy Fuels Corp to buy 170 million-gallon equivalents of RNG through 2026 in support of its commitment to reduce its greenhouse gas emissions 12% by 2025. RNG biogases used for transportation fuel are refined into an ultra-low-carbon, low-emissions source of energy, especially when made from food waste processed in anaerobic digesters, resulting in a net carbon-negative over its lifecycle.
- *Newlight Technologies*: This LLC has patented greenhouse gas-to-plastic technology that makes plastics from greenhouse gases captured from biological waste. Most plastic is still made from petroleum, a fossil fuel. However, the biodegradability of this new form of plastic needs to be further assessed.
- *MOVE Systems*: This is a company that has made hybrid food truck vehicles that use renewable energy. The company has helped brands like Dunkin, Nathan's Famous, and others reach beyond their stores in a sustainable way.

---

[8] https://energy-vision.org/wp-content/uploads/2019/10/EVNews-Winter-2016.pdf, Energy Vision News, Quarterly © 2016 Energy Vision Inc.

Living off the Grid: The New Normal

More and more people worldwide are living off the grid.[9] That is grid as in *power grid*. Instead of getting electricity from public utilities that draw on regional transmission systems, many consumers are seeking out sustainable sources and alternative energy plans. The idea of freedom from costly utility bills has always been the dream of self-reliant buffs. However, now it is not just a few who want that freedom. Since most public utilities produce energy with fossil fuels and thus have high carbon footprints, many people want to cut ties and go their own, more environmentally responsible way.

For a few decades, the trend of living off the grid has been largely driven by the efforts of individuals, families, and often rural communities across the United States. According to *Home Power Magazine*,[10] at least 180,000 families are living off the grid in the U.S., and that number increases each year. Some communities have banded together to go off the grid; not surprisingly, they differ in terms of goals, geographies, and modus operandi.

- *Three Rivers Recreation Area*: Located in Central Oregon, this arid, rustic area is home to between 400 to 600 people of varied backgrounds and means. The self-sufficient homes use solar-powered electricity for high-speed internet, satellite television, and appliances.

- *Greater World Community*: The sculptural, art-filled homes in this community in Taos, New Mexico, are innovative and sometimes fanciful in design. Each one has its own greenhouse, propane stove, and water-recycling system. These homesteading "earthships" are built by a construction company named Earthship Biotecture, with an emphasis on sustainability. All are solar- or wind-powered, with numerous energy-saving features. They are made of locally sourced mud plaster, clay, and recycled materials such as tires and bottles. Visitors can tour the community and rent a place to stay.

---

[9] https://science.howstuffworks.com/environmental/green-science/living-off-the-grid.htm, "How Living Off the Grid Works," Charles W. Bryant, *howstuffworks*.

[10] https://www.homepower.com/home, Home Power.

- *Breitenbush*: Also in Oregon, this community has about 85 residents who live on 154 acres in a mountain setting. Their organization offers personal sanctuary retreats to guests and attracts many visitors to its hot springs and conference center. A nearby river and geothermal wells provide the energy for the community, which shares kitchens and avoids unnecessary appliances. Building materials are made, in part, from trees that have blown down in the property's forest.

- *Earthaven*: Located on 329 acres near Asheville, North Carolina, the planned community of Earthaven Ecovillage is home to about 100 residents of all ages. It uses solar panels and hydropower to meet its energy needs. Its houses and small businesses are set in an environment composed of forests, mountains, wetlands, and flood plains. Its orchards and gardens use permaculture techniques, and its buildings are designed to be energy-efficient, natural, and sustainable.

- *Emerald Earth*: This smaller intentional community of up to 20 people in Mendocino County, California, is housed in several off-the-grid buildings, surrounded by almost 200 acres of meadows and forests. In this relatively remote, rural place, residents grow the bulk of their own food. They are guided by the idea of benefiting, not depleting the Earth. The community provides for its own power needs and uses natural building methods and materials, such as straw, clay, and wood. The community practices conservation, permaculture, composting, and organic gardening and gives workshops on these topics.

- *Dancing Rabbit Ecovillage*: In rural northeast Missouri, approximately 40 to 70 residents live on a 280-acre site. Their commitment to environmental sustainability involves generating renewable energy with solar panels and wind turbines, "harvesting" rainwater, growing food organically, and composting. They share a limited number of cars, implement permaculture in all their gardens, and build exclusively with green construction designs and methods, using locally found or reclaimed lumber.

These communities, with their range of sizes and locales, are by no means the only places that focus on living a sustainable lifestyle. Others take a different approach to the concept of "off the grid." In Waltham,

Vermont, the rooftops in a low-income development brandish shiny, solar panels. These panels take in the sun's rays, which feed energy to backup batteries in the basements of the development. However, in this case, it is not single homeowners nor an intentional community driving the sustainability initiative. It is an innovative program from the local electric company.

The *New York Times* describes how the electric company, Green Mountain Power, lets their Vermont customers disconnect from the grid and turn their homes into mini power plants.[11] In a clear break from business as usual, Green Mountain Power is realizing both environmental and financial advantages. The less electricity Green Mountain Power gets from the regional transmission system, the lower the fees it pays. If homeowners are producing excess electricity, Green Mountain Power can remotely draw it from customer's batteries for use elsewhere.

In addition to offering their customers this chance to live off the grid, Green Mountain Power has also offered them access to Tesla's Powerwall home battery system, which was released in 2015. According to the *New York Times*, Green Mountain Power is starting a new program offering the Tesla battery to as many as 2,000 customers for $15 a month over 10 years, or a one-time payment of $1,500. Green Mountain Power is a leading example of the transformation the utility sector is undergoing.

The paradigm is shifting. No longer is the model to have one big distant, air-polluting, CO_2-emitting power plant delivering electricity over a massive, aging, inefficient grid to many customers at a fixed, regulated price. Vermont is nimbly embracing a modern system with multiple power sources including solar farms, rooftop solar panels, and wind turbines, coupled with batteries. These sources have lower production and transmission costs, lower greenhouse gas emissions, and variable pricing keyed to demand.

---

[11] https://www.nytimes.com/2017/07/29/business/energy-environment/vermont-green-mountain-power-grid.html?mwrsm=amp-email, "Utility Helps Wean Vermonters from the Electricity Grid," Diane Cardwell, *New York Times*, July 29, 2017.

What Does Wind Power Have? Jobs, Jobs, Jobs!

In 2017, the fast-growing wind power industry employed 102,000 workers in the United States, which was 25,000 more workers than the previous year. The American Wind Energy Association (AWEA)[12] reported that the U.S. Department of Energy (DOE) verified that the number of Americans employed by the wind power industry is more than those working at nuclear, natural gas, coal, or hydroelectric power plants. Working on wind turbines is turning out to be a fast-growing job sector in the U.S., now and in the foreseeable future. According to DOE's Wind Vision Report, 380,000 American wind jobs could be created by 2030.[13,14]

By 2017, wind power energy had doubled its energy output over the previous five years. It is now the country's third-largest source of electric capacity behind fossil fuels and nuclear energy. Wind power produces about four times more American electricity than solar, and its energy output is ahead of hydroelectric. Wind is now 8.4% of total U.S. energy. This newer, cheaper, safer, and more-efficient paradigm is being implemented so that home-owners can collect their own energy via solar panels and wind turbines, selling the excess back to power companies.

Bury the Power Grid

It ought to be clear to everyone that the climate and the oceans are warming, which means that there are and will be more frequent intensive storms. Since 2005, mega-storms have devastated New Orleans (Hurricane Katrina), New York and New Jersey (Superstorm Sandy), Houston (Harvey), and Miami (Irma). In all cases, flooding from rain and storm surges caused widespread devastation, and millions lost power for days or

---

[12] http://www.awea.org/, American Clean Power Association.

[13] https://www.energy.gov/eere/wind/wind-vision, Wind Vision, Office of Energy Efficiency & Renewable Energy.

[14] https://www.energy.gov/articles/doe-releases-second-annual-national-energy-employment-analysis-0, "DOE Releases Second Annual National Energy Employment Analysis," Department of Energy, January 13, 2017.

weeks from downed power lines. Will your city be next? What can be done to prepare for the next big blow?

One obvious step is to bury the power grid. The Climate Institute's North American Supergrid Initiative is proposing that bold yet visionary step.[15,16] The country's electric grid is old infrastructure; some parts of it are over 130 years old. The first electric grid dates back to 1882 when Thomas Edison opened the country's first power plant in lower Manhattan. Today's grid is increasingly obsolete and vulnerable. It makes good economic sense to replace vulnerable old copper wire grids with buried optical cable ones.[17] Doing so will be costly, and the ability to take on such massive new infrastructure projects will vary greatly among the richer and poorer nations of the world. However, think of the cost of millions of households and businesses without power for days or weeks during the increasingly severe storms forecast with climate change.[18] The next big storm will happen somewhere soon.

Shell Maps Radical Plan for Energy: Hydrogen

Unlike other large oil firms, Royal Dutch Shell sees hydrogen gas providing 10% of global energy consumption by 2070.[19] It envisions many cars, trucks, and even airplanes running on hydrogen, despite the current enthusiasm for battery-electric vehicles. Hydrogen is a fuel that does not emit greenhouse gases. It can be used in fuel cells or internal combustion engines, which are being developed for passenger cars. Fuel-cell powered buses have been on the road for many years, proving hydrogen's viability.

---

[15] http://climate.org/, Climate Institute.

[16] http://northamericansupergrid.org/, North American Supergrid.

[17] https://futurism.com/should-the-us-put-power-lines-underground, "Should the US put power lines Underground?: What are customers willing to pay for to ensure reliability and mitigate risk?", *Futurism*, September 15, 2017.

[18] https://www.cfr.org/backgrounder/how-does-us-power-grid-work, "How Does the US Power Grid Work?" James McBride and Anshu Siripurapu, *Council on Foreign Relations*, May 14, 2021.

[19] https://www.shell.com/energy-and-innovation/the-energy-future/scenarios.html, Shell Scenarios.

Now, Shell is being pushed by activist shareholders to be more transparent and more environmentally responsible. The Anglo-Dutch firm plans to convert some of its natural gas plants to produce hydrogen.

Shell sees the demand for oil stagnating in the 2020s, with natural gas declining in the 2040s, as governments mandate that many grids be powered 100% by wind, solar, and hydro. All the measures it foresees in its company scenario will limit global temperature warming to 1.7° Celsius to 1.8° Celsius (3.1 to 3.2° Fahrenheit), which is under the Paris Agreement target of 2° Celsius (3.6° Fahrenheit), but above the aspirational target of 1.5° Celsius (2.7° Fahrenheit).[20] Shell acknowledges that meeting the tougher target would require governments to tax or price carbon by 2030.

So, while Shell is relatively more enlightened than its peers and recognizes that governments must complement energy industry efforts, the company is vague on how it plans to achieve its ambitious targets, and it is silent on how to encourage its peers to match or exceed those targets. It is also silent on how to pressure governments to do what it says they must do — price carbon and curtail carbon subsidies. How serious Shell is about implementing its corporate vision remains to be seen.

Net Zero by 2050 Is Feasible

Researchers at Stanford University have published a feasibility study conducted to determine what it would take to convert 143 countries worldwide to 100% zero-carbon energy wind, solar, and storage systems by 2050.[21] They say it could be done by 2030 but for a variety of reasons have

---

[20] https://unfccc.int/process-and-meetings/the-paris-agreement/the-paris-agreement, The Paris Agreement.

[21] Jacobson, M.Z., M.A. Delucchi, M.A. Cameron, *et al.*, Timeline and Land Area Required to Transition the All-Purpose End-Use Power of 143 Countries to 100 Percent Wind-Water-Solar; Five Reasons End-Use Demand Decreases 57.1 Percent Along the Way in Impacts of Green New Deal Energy Plans on Grid Stability, Costs, Jobs, Health, and Climate in 143 countries. *One Earth*, 1, 449–463, (2019), doi: 10.1016/j. oneear.2019.12.003; and https://web.stanford.edu/group/efmh/jacobson/Articles/I/WWS-50-USState-plans.html

chosen the later date. They estimate such a system would cost $73 trillion to implement but would recoup the entire cost in 7 years.[22] According to the Stanford researchers, building the new system would incur a carbon emission footprint less than 1% of the current energy system. It would create millions of jobs and save many of the 7 million lives lost yearly to pollution.

However, they devote little attention to funding sources, implementation mechanisms, or ways to overcome political and corporate resistance to their ambitious plan. They do provide a country-by-country breakdown of how it could be achieved.

One has to admire the audacity of the scientists' net-zero carbon emissions vision. Its scope is simply stunning.

Denmark Surges in Renewable Electricity

In 2019, Denmark produced over half its energy from renewables, mostly from wind, only a little from solar.[23] The year 2019 saw a big jump from 2018, when 41% came from renewables. The increase was largely due to the coming on line of the Horns Rev 3 407-megawatt offshore wind project. Two other projects authorized by the Danish parliament in 2018 are upping Denmark's wind energy output even more. In addition, parliament passed another even bigger climate act. It calls for building one or more islands around which will be placed offshore wind turbines capable of generating 10 gigawatts of power, enough to provide electricity for 10 million homes. Denmark is committed to cutting greenhouse gases 70% by 2030 and reaching net zero emissions by 2050.

---

[22] https://www.sciencealert.com/stanford-researchers-have-a-plan-to-tackle-the-climate-emergency/amp "Stanford Researchers have an exciting plan to tackle the climate emergency worldwide," Tessa Koumoundourus, *Science Alert*, December 27, 2019.

[23] https://cleantechnica.com/2020/01/05/denmark-passes-magic-50-in-renewable-electricity-generation-milestone/ "Denmark passes magic 50% in renewable electricity generation milestone," Jesper Berggreen, *Clean Technica*, January 5, 2020.

Further, Denmark will decarbonize all sectors, including energy, transportation, and agriculture. It will cost the equivalent of between $29 billion and $44 billion dollars, coming mostly from private sources. The act aims not for the possible, but for the necessary. Denmark's task is to make the necessary possible.

Denmark hopes other countries will follow its lead. Norway already gets 98% of its energy from renewables, mostly hydro power. In 2015, Germany got 30% of its energy from renewables, but its progress has slowed, in part by Germans' antipathy to having wind turbines near their homes. Grid limitations also inhibit delivery of power from where it is generated in the north of the country to where industry needs it in the south. Eastern European countries face difficult challenges in generating more energy from renewables due mostly to their large legacy coal industries. Still, the European Union has proposed that all its members acknowledge that a climate emergency exists and commit to neutralizing carbon emissions by 2050.

Nuclear Energy — Where Are We Today?

Nuclear energy has long been touted as the ultimate fuel source, limitless and emission-free. However, the term "nuclear energy" conflates two very different processes and technologies: *fission* and *fusion*. For anyone interested in or concerned about the climate, getting acquainted with the facts, pros, cons, and issues surrounding both varieties of nuclear energy is a good idea.

Nuclear energy, the once-promising alternative, emits no greenhouse gases, but it presents a challenge because fission's fuel is not renewable. In fact, its spent fuel takes thousands of years to lose its lethal radioactivity. Fusion nuclear may someday have a role in a post-fossil fuel world, but that day is also decades away.[24]

---

[24] https://www.world-nuclear.org/information-library/current-and-future-generation/nuclear-fusion-power.aspx "Nuclear Fusion Power," World Nuclear Association, June 2021.

Nuclear fission is the splitting or breaking apart of the radioactive atomic nuclei of the heavy metal elements uranium or plutonium. Allied scientists learned how to split atoms in the secretive Manhattan Project during World War II. Their work culminated in the explosion of nuclear bombs at Hiroshima and Nagasaki in Japan in 1945.

Nuclear fission is problematic for several reasons. Although a fission reactor produces no carbon emissions, it produces spent fuel that is radioactive for tens of thousands of years, with radiation that is deadly to humans. While the engineering problem is not of the same order of difficulty as creating a sun on Earth, ensuring the safe storage of the radioactive waste for millennia is not a trivial problem. It has yet to be solved. Nobody wants it anywhere near them, for understandable reasons. The problem has its own acronym, NIMBY, which stands for "Not In My Back Yard."

The nuclear fission plants built in the 1950s and 1960s produced a lot of waste. Some is stored in caves, some deep underground. Some has been lost — nobody knows for sure where or how it is stored. And we are 100 years into a multi-millennial disposal cycle. Germany wants to close all its fission reactors from the 1960s, but it has nowhere to stash the spent but still radioactive fuel rods. Nobody wants them, including the Germans.

Nuclear diehards have designed new, small, fail-safe reactors that generate far less waste than old designs. However, fans of fission have not been able to solve the problems of how and where to store fission's radioactive wastes for thousands of years. Nor have they resolved how to avoid the regulatory hurdles fission must pass to be licensed and the insurance costs for protecting against a Chernobyl-like disaster. While energy produced by nuclear fission is relatively inexpensive, the design, licensing, insurance, construction, and operating costs simply make it uneconomical given current alternatives.

Furthermore, everyone is aware of the nuclear accidents that have befallen supposedly fail-safe plants — Three Mile Island, Chernobyl, and Fukushima Daiichi. Lessons have been learned, and today's designers assure us that the new designs cooled by liquid sodium won't evaporate causing a meltdown if power to the cooling system were to be interrupted. Still, would you want one located near your home?

Bill Gates, the multi-billionaire founder of Microsoft, has invested heavily in the start-up nuclear fission company Terra Power.[25] He sees his company providing significant quantities of the carbon emission-free energy the world will need in 2050. Gates sees reaching zero emissions as the way to overcome the climate crisis, and he presumes that nuclear technology will take us there.

The second process for creating energy from atomic reactions is *nuclear fusion*, the merging of two hydrogen atoms into one helium atom. This is the same process that occurs within our sun and all other stars. Both splitting and fusing atoms require — and release — great quantities of energy. Nuclear scientists from several countries are working hard to achieve a fusion reaction that produces more energy than it takes to make it happen, the equivalent of creating a mini-sun on Earth.

Nuclear fusion does not have fission's waste disposal problems, but it does share all its other negatives. Operational nuclear fusion, despite recent design advances, is still many decades away. By then, the world will very likely be run on wind, solar, and hydrogen. Fusion will always be an expensive niche energy provider.

Despite decades of work and billions of dollars invested in the goal, no viable fusion reactor yet exists. Given enough time, money, and human ingenuity, one day nuclear fusion will power a percentage of the world's energy. But that day is distant. As some have predicted, fusion may play an important role by 2100, if the human race has met the challenge of preserving a livable Earth.

So ClimateYou believes that neither nuclear fission nor fusion will become our civilization's dominant form of energy. Abundant nuclear energy may facilitate the long-term survival of the human race, but it will be far from the only, or even the most important, determinant of our future.

---

[25] https://www.cnbc.com/2021/04/08/bill-gates-terrapower-is-building-next-generation-nuclear-power.html "How Bill Gates' company Terra Power is building next-generation nuclear power," Catherine Clifford, *CNBC*, April 8, 2021.

Transportation

San Francisco's Golden Gate Bridge (California, 2017)

Hydrogen Fuel Cell Vehicles Become a Reality

Researchers have recently developed ways to produce hydrogen from water cheaply. They use common metallic catalysts rather than expensive rare ones, in addition to relying on renewable energy to break water into its component parts without generating greenhouse gas emissions. The hydrogen fuel cells' only by-product is water. So, while the next transition in transportation — electric vehicles (EVs) replacing carbon emission-spewing internal combustion engine (ICE) vehicles — is well under way, within a decade or two, hydrogen-fueled vehicles will likely be on the road as well.[26]

This scientific breakthrough will ease the problems of producing cars that run on hydrogen fuel cells. The new development is a mechanism that makes hydrogen fuel from water using a new biomaterial that enables water to be separated into hydrogen and oxygen. The invention then mixes hydrogen with oxygen and generates electricity to power car motors that release water vapor, not toxic fumes produced by gasoline.[27] Cars and trucks running on hydrogen fuel cells are considered to be the most environmentally clean type of vehicle.

According to the Union of Concerned Scientists, fuel cell cars and trucks can cut greenhouse gas emissions by over 30%. Honda, Hyundai, and Toyota are early producers of hydrogen fuel cell cars.

Free Two-Day Shipping and Climate Change

What do free two-day shipping offers have to do with climate change? Quite a lot, actually. They result in more partial-load delivery trucks on the road,

---

[26] Jordan, P., Patterson, D., Saboda, K. *et al.* Self-assembling biomolecular catalysts for hydrogen production. *Nature Chem* 8, 179–185 (2016). https://doi.org/10.1038/nchem.2416

[27] https://www.ucsusa.org/resources/how-clean-are-hydrogen-fuel-cell-vehicles#. Vp2h2iorLIU "How clean are hydrogen fuel cells?", Union of Concerned Scientists, October 9, 2014.

spew greenhouse gas emissions, and increase traffic congestion. In a report titled "Why You Should Think Twice Before Opting for Free 2-Day Shipping" published on *HuffPost*,[28] Laura Paddison points to a multi-year study by the University of Delaware that looked at how online shopping in Newark, Delaware, was having a negative impact on the transportation system, especially for emissions. Delivery trucks catering to quick two-day deliveries are unsustainable and inefficient, putting many more trucks on the road and increasing the amount of greenhouse gases emitted.

In some cases, ordering online can be both more convenient and better for the environment, when measured against individual customers driving to stores and back. However, the convenience may be overshadowing some long-term environmental consequences.[29] To get a package you ordered to you on time, a company may have to ship a purchase from a distant location, sometimes by air, which also increases emissions. Quick, free shipping — while it may save you a trip to the store in your car, with its attendant downsides of time, emissions, and road wear — may contribute to induced demand and over-consumption, which are at the root of the human toll on the environment.

The Global Shipping Industry and Blockchain

What is blockchain? Blockchain technology is a list of digital records, or blocks, that are linked together using protected computer codes.

In a *Bloomberg* news story titled "The Blockchain on the High Seas," Matt Levine looks at how the blockchain accounting system can revolutionize the global shipping industry.[30] Levine points out that most shipping

---

[28] https://www.huffingtonpost.com/entry/2-day-shipping-environment_us_5a0e1374e4b04 5cf43706864?euq, "Why You Should Think Twice Before Opting for Free 2-Day Shipping," Laura Paddison, *Huffpost*, December 6, 2017.

[29] http://www1.udel.edu/udaily/2016/feb/online-shopping-020516.html "Shop Right: Online shopping might not be as green as people think it is," Diane Kukich, *UDaily*, February 5, 2016.

[30] https://www.bloomberg.com/view/articles/2018-04-22/shipping-industry-has-a-block-chain-problem "The Blockchain on the High Seas," Matt Levine, *Bloomberg*, April 22, 2018.

companies use their own blockchain platforms, or websites, to accomplish real-time cargo tracking. However, shipping companies would greatly benefit from an industry-wide consortium that shares a unified database via blockchain technology.

Shipping is currently responsible for about 3% of global greenhouse gas emissions. Unconstrained, however, emissions from shipping are projected to reach 10% of total global emissions by 2050, as industry emissions rise. If the shipping industry can adopt a unified blockchain accounting system to track all shipments, the impact could be significant — bringing down costs, delivery and settlement times, and back-office employment. Making shipping more efficient will reduce the overall need for fossil fuel. An agreement reached through the auspices of the International Maritime Organization commits the shipping industry to cut its emissions by 50% by 2050.[31]

The problem is getting all shipping companies to agree to adopt one system, rather than each company maintaining its own. Nevertheless, an industry-wide compact for all to use just one blockchain accounting system would benefit all. It would work by reducing both development and maintenance costs, yet at the same time assuaging each company's concerns about the safety and security of their proprietary data. There is at least a fair chance that one common system can be negotiated.

Together, agreements on emissions and digitalized secure accounting will revolutionize the shipping industry, boost the global economy, and contribute to achievement of the United Nations Framework Convention on Climate Change Paris Agreement.[32]

Americans Need to Drive Less

One of America's biggest source of carbon emissions is people driving personal automobiles (except during the coronavirus pandemic). What can be done to change this behavior? Telling Americans they have to drive

---

[31] https://www.imo.org/en/MediaCentre/PressBriefings/Pages/06GHGinitialstrategy.aspx, "UN body adopts climate change strategy for shipping," International Maritime Organization, April 13, 2018.
[32] https://unfccc.int/process-and-meetings/the-paris-agreement/the-paris-agreement, The Paris Agreement.

less is a non-starter. In most places, people have no viable alternative for getting to jobs, grocery stores, appointments, and so on. Disinvestment in public transit for decades has left the "doing away with the car" option so unattractive or inconvenient as to be moot.[33]

There are pros and cons of adopting alternative ways to cut U.S. transportation emissions significantly. Providing cash for old cars that spew a lot of greenhouse gas emissions is an expensive way to make far too slow an impact on emissions, given that only about 6% of Americans buy a new car each year. European cities have implemented restrictions on cars in urban areas, and New York City is planning to do the same in the near future. When the European cities have put driving restrictions in place, initial opposition has quickly dissipated.

Setting more stringent Corporate Average Fuel Economy (CAFE) standards is one way to reduce emissions through regulations that require increases in the fuel economy of cars and light trucks. California has had a special status with the ability to set its own emission standards because of its heavy reliance on automobile transport. The automobile industry has adhered to these standards for many years, which for cost and efficiency reasons have become the de facto U.S. national standards.

Simply put, Americans need to drive less, which may impinge on where they live. Before the COVID-19 pandemic, there was a nascent movement to return to the denser living of the cities, rather than to commute to work from sprawling suburbs. For most people this is a hard sell, given negative perceptions of urban life. Since the coronavirus pandemic, some people are opting to move away from fast-paced, expensive urban centers and to work remotely from smaller, more congenial towns, where living is easier. One idea is to decentralize workplaces to reduce commuting time. Implementing widely high-tech virtual conference software would enable more people to work from home (as has been done during the coronavirus outbreak in 2020–2022). These transportation emissions issues and ways to address them are key to achieving net zero emissions by 2050.

---

[33] https://www.huffpost.com/entry/the-democrats-baffling-blind-spot-on-carbon-emissions_n_5da4db19e4b0cad669a9b705 "Democrats' Baffling Blind Spot on Cars," Michael Hobbes, *Huffpost*, November 4, 2019.

Innovation and Technology

Wind power turbine (Maine, 2011)

Solar Plane a Success

Flying an airplane fueled only by the sun has now been done with great success. In April 2016, the solar-powered plane named Solar Impulse 2 landed in California after flying over the Pacific Ocean, a route that precluded emergency landings.[34] The 62-hour non-stop flight began in Kalaeloa, Hawaii, on April 21 and covered almost 2,500 miles.

Solar Impulse 2 began its history-making flight around the world in March 2015, when pilots Bertrand Piccard and Andre Borschberg took off from Abu Dhabi. The plane has made stops in Oman, Myanmar, China, Japan, and Hawaii. About 17,000 solar cells cover its expansive wings — longer than those on a Boeing 747 — and provided enough power for the propellers and for charging the plane's batteries. The ideal flight speed of the Solar Impulse 2 was about 28 miles per hour, which can double during the day when the sun's rays are strongest. The carbon-fiber aircraft weighs more than 5,000 pounds, or about as much as a mid-size truck.

According to the Environmental Protection Agency, the flights of aircraft account for about 11% of all U.S. transportation greenhouse gas (GHG) emissions and more than 3% of total U.S. GHG emissions.[35] The European Commission has warned that by 2020, international aviation emissions could be 70% higher than in 2005, even if fuel efficiency improves by 2% a year, due to increased travel.[36]

Solar Impulse 2 did a record-breaking solo flight for five days and nights without fuel, when it flew from Nagoya to Hawaii. It made more stops in the U.S. before crossing the Atlantic Ocean.

The pilots kept a log of their journeys that showed the plane's altitude, speed, and navigation. One entry said that by attempting the first solar flight around the world, they were "pushing back the boundaries of the possible, going into the unknown, and taking on a project deemed impossible by industry experts." The pilots emphasized that conservation of resources and the pursuit of new technological solutions to energy usage

---

[34] http://www.solarimpulse.com, Solar Impulse.

[35] https://www.epa.gov

[36] http://ec.europa.eu/clima/policies/transport/aviation/index_en.htm, European Commission: Climate Action.

constitute "the new adventure of the 21st century." During the flight, the pilot talked by satellite to the Secretary General of the UN upon the momentous signing of the Paris Agreement.[37]

Waste-to-Fuel Revolution

The use of renewable natural gas (RNG) is seeing a dramatic uptick by businesses and municipalities throughout the United States. Renewable natural gas is made by breaking down organic material found in wastewater, agricultural biomass, food scraps, and yard residues to produce biogas. The process uses anaerobic digesters to break down the organic material, and then the resulting slurry is refined into an ultra-low-carbon, low-emissions source of energy and transportation fuel, all with a fraction of the carbon footprint of diesel, gasoline, or other petroleum-based fuels.

In New York City, almost four million tons of organic waste is generated per year, which, if converted to RNG, could fuel the city's entire heavy-duty fleet of trucks. The Newtown Creek wastewater plant in Brooklyn, the largest of New York City's 14 wastewater treatment plants, features eight new RNG anaerobic digesters that resemble large, futuristic stainless steel-clad eggs. These structures process as much as 1.5 million gallons of sludge every day.[38]

Nationally, the U.S. Environmental Protection Agency (EPA) has included RNG in its Renewable Fuel Standard (RFS2),[39] which has been pivotal in driving development and deployment of RNG as a net carbon-negative transportation fuel. Vancouver in British Columbia announced it would phase out non-renewable natural gas by 2050[40] and work to use

---

[37] https://unfccc.int/process-and-meetings/the-paris-agreement/the-paris-agreement, The Paris Agreement.

[38] http://www.nyc.gov/html/dep/html/environmental_education/newtown_wwtp.shtml, Education Programs, NYC Environmental Protection.

[39] https://www.epa.gov/renewable-fuel-standard-program/renewable-fuel-standard-rfs2-final-rule, "Renewable Fuel Standard (RFS2): Final Rule," United States Environmental Protection Agency.

[40] https://globalnews.ca/news/2958288/city-of-vancouver-votes-to-ban-natural-gas-by-2050/ "City of Vancouver votes to ban natural gas by 2050," Jill Slattery, *Global News*, September 22, 2016.

more RNG in its trucks. London has announced it would stop buying or leasing diesel trucks for its fleets,[41] and it has received permission to generate RNG at the city landfill. It has advanced its ban on the sale of new gas diesel and hybrid vehicles to 2035 from 2040.

A *New York Times* op-ed[42] argued that NYC municipal fleets should stop buying diesels and start buying RNG trucks now in order to meet the city's ambitious emissions goals. The co-authors Robert Catell and Joanna Underwood wrote that the New York City administration of Mayor Bill de Blasio "has set clear, aggressive goals for reducing greenhouse gas emissions that would make the city a national leader in mitigating climate change."

Vacuum Up That Carbon

A good phrase to know: *Direct Air Capture*. It is a process being developed that does what trees do — it takes carbon dioxide out of the air. Direct air capture is a radical variant of carbon capture and storage (CCS). Several start-up companies are pursuing this technology, and all are out of the lab and into the pilot phase of development.

In the digital news outlet *Quartz*, Akshat Rathi[43] reported that companies piloting the direct CO_2 capture are Switzerland's Climeworks,[44] Canada's Carbon Engineering, and the U.S.'s Global Thermostat. All are building machines that, at reasonable costs, can capture CO_2 directly from the air.

In the spring of 2017, Climeworks set up its first commercial unit near Zurich, Switzerland, that can capture about 1,000 metric tons of CO_2 from

---

[41] http://www.standard.co.uk/news/london/city-of-london-to-stop-buying-diesel-vehicles-in-boost-for-pollution-battle-a3307421.html "City of London to stop buying diesel vehicles in boost for pollution battle," Nicholas Cecil, *Evening Standard*, July 29, 2016.

[42] http://www.nytimes.com/2016/08/19/opinion/how-garbage-trucks-can-drive-a-green-future.html, "How a garbage truck can drive a green future," Robert B. Catell and Joanna D. Underwood, *New York Times*, August 19, 2016.

[43] https://qz.com/author/akshatqz/, Akshat Rathi, *Quartz*.

[44] http://www.climeworks.com/, Climeworks.

the air each year (equivalent to 20 U.S. households' annual emissions).[45] According to Rahti's article, "the captured CO_2 is supplied to a nearby greenhouse, where a high concentration of the gas boosts crop yield by 20%."

This innovative technology is projecting price targets at a scale that imply viability. If their early success generates more investor interest, both the number of business ventures and the pace of technological breakthroughs should accelerate. This quickening of the pace of change is important. Reducing carbon emissions alone may not proceed fast enough and go far enough to avoid catastrophic climate change. Removing significant amounts of CO_2 from the air can help deter or delay the worst consequences.

Geoengineering Is Not the Answer

It seems clear that geoengineering is not the answer to climate change. Geoengineering is a catch-all term for proposed technological methods of large-scale intervention in the climate, such as reducing the amount of solar radiation that reaches the earth via injection of reflective aerosols or iron fertilization of the ocean to absorb CO_2 from the atmosphere. Thinking that such technological fixes will stop global climate warming is a fantasy. Technology does not and cannot address the underlying causes of climate change, which are capitalism, civilization, and the notion of progress. Long-term sustainability of the Earth as a viable habitat for humans and other life forms depends on responsible stewardship of the resources of the Earth.[46]

We cannot keep depleting those resources indefinitely. They are vast, but not inexhaustible. We cannot only consume; we must also replenish. We cannot forever pollute the air, land, and ocean. The Earth's biosystem is

---

[45] http://www.sciencemag.org/news/2017/06/switzerland-giant-new-machine-sucking-carbon-directly-air, "In Switzerland, a giant new machine is sucking carbon directly from the air," Christa Marshall, *Science*, June 1, 2017.

[46] https://www.huffpost.com/entry/geoengineering-climate-change_n_5ae07919e4b061c0 bfa3e794, "The dangerous belief that extreme technology will fix climate change," Aleszu Bajak, *Huffpost*, April 27, 2018.

complex. While it has the ability to regenerate at least some of its resources, these processes, such as soil formation and the biodiversity of an ecosystem, often take hundreds or thousands of years. Yet, we can help or hinder that capacity.

Thomas Hobbes once characterized the life of man as "poor, nasty, brutish, and short." Yet humans have, over the centuries, improved their lot. While many still live in poverty, unsure of their next meal, more and more humans have achieved living conditions with at least minimal levels of food security, clothing, and shelter from the elements. And many more people live lives of comfort, security, and opportunity undreamed of by our ancestors. Can such munificence be maintained and expanded to include all of humanity without causing a collapse in the carrying capacity of the Earth? Climate scientists say yes, if we stop burning fossil fuels, which spew into the atmosphere billions of tons of greenhouse gases that trap heat and cause global temperatures to rise.

The sun and the wind provide much more energy than we need to maintain our current or any anticipated lifestyles. We do not need carbon-based energy. Effecting an energy transition will not be easy, especially with the fossil fuel industry as large, powerful, and pervasive as it is. Curbing its power means challenging some of the basic tenets of capitalism, exerting uncommon amounts of political will, and instilling a new ethos of sustainability in all of humanity — by tomorrow or the day after, certainly by mid-century.

We do not have much time before the bill for our profligate over-spending of the Earth's vast resources comes due. The cost will be intolerable pollution, droughts, heatwaves, floods, storms, famines, mass migrations, and political unrest. Increasingly, these negative effects will befall us if we do not act decisively now to curb carbon emissions. Geoengineering alone will not save us, but a combination of politics, economics, and sociology, together with appropriate technology can, and must.

Let's Capture Methane

Climate change science is without question one of the most fascinating and exciting fields to be in today. For example, MIT's *Technology Review* has an article by James Temple[47] describing how scientists have created a new molecular system that could capture a few billion tons of methane from the atmosphere to reduce short-term global warming. Methane traps over 80 times more heat during the first two decades in the atmosphere than does carbon dioxide.

One promising family of materials for trapping CH_4 is called zeolites. Zeolites are porous systems that can concentrate methane in industrial applications because they have absorption capacities that selectively bind methane and carbon dioxide (CH_4/CO_2) and methane and nitrogen (CH_4/N_2). The scientists recommend that zeolites should be evaluated for their potential to reduce methane concentrations in the atmosphere from the current level of approximately 1,860 parts per billion (ppb) to pre-industrial levels of about 750 ppb.

The main idea discussed in the paper by Jackson *et al.* titled "Methane removal and atmospheric restoration"[48] seems radical at first, but it has a firm rationale behind it. Instead of capturing carbon dioxide (of which there is a great deal), let us capture methane (of which there is relatively little), because we would get a lot bigger bang for our buck. Capturing methane could reduce the heat-trapping effect of the atmosphere to a significantly greater degree in the short term. Thus, capturing methane could slow the heating of our climate and give us more time to cut carbon emissions to zero.

Energy System Disruptors: Pros and Cons

Oceanic methane hydrates — ever heard of them? They may be about to disrupt the global energy system in a potentially huge way.[49] The resources

---

[47] https://www.technologyreview.com/s/613556/turning-one-greenhouse-gas-into-another-could-combat-climate-change/ "Turning one greenhouse gas into another could combat climate change," James Temple, *Technology Review*, May 20, 2019.

[48] Jackson, R.B., E.I. Solomon, J.G. Canadell, M. Cargnello, and C.B. Field. 2019. Methane removal and atmospheric restoration. *Nature Sustainability* 2: 436–438. www.nature.com/natsustain; and https://www.nature.com/articles/s41893-019-0299-x

[49] https://foreignpolicy.com/2020/01/09/fracking-oceanic-methane-hydrates-global-energy-landscape-bonanza/ "The World's Next Energy Bonanza," Alex Gilbert, Morgan D. Bazilian, and Sterling Loza, *Foreign Policy*, January 9, 2020.

are widespread, off many coasts. The race to commercialize them is on, with winners expected within five years.

However, several factors stand in the way of their full development as an energy source. First, there is a near-total lack of regulation. Second, methane is a volatile gas. A major leak or an explosion would have major environmental impact. Third, high costs could be dampening factor, although the cost of commercializing methane hydrates is still unknown. If methane hydrates cannot beat the ever-diminishing costs of wind and solar plus storage, their future is limited, at least initially.

It is also important to consider that methane is a carbon-based fossil fuel. Its combustion produces CO_2, the major driver of climate change. The world is currently trying to wean itself off fossil fuels by replacing them with renewable energy resources such as wind and solar. The transition is going slowly, but it must accelerate to reach a 45% reduction by 2030 and achieve a target of close to zero carbon emissions by 2050. These goals must be met, if the world is to avoid some challenging, or even deadly, effects of global heating.

If there is to be another global energy disruptor, it is most likely to be hydrogen fuel cells (HFCs). HFCs, like batteries, store energy. They have an advantage over standard lithium-ion batteries as a fuel source because they can store energy for days or weeks, not hours. The hydrogen comes from splitting water into its component parts, hydrogen and oxygen. It takes a fair amount of electricity to break apart the water atoms. The process until recently has been costly and high in carbon emissions, as the electricity needed was generated by burning fossil fuels.

However, the electric power can now come from wind or solar, reducing both cost and GHGs. Even so, HFCs have been more expensive than batteries because they relied on costly rare metals to catalyze the breakup of water. Yet, researchers in Australia have accomplished the same results with cheap, common catalysts. Researchers have a few more hurdles to surmount, but HFC technology is promising both for vehicles and for reserve power for renewable energy sources that have intermittency issues.

Other potential global energy disruptors are nuclear fission and fusion. While some hope that GHG emission-free nuclear fission (the current form of nuclear energy) might play the role of the global energy disruptor, the window for nuclear fission has probably already closed for cost and radioactive waste disposal reasons. The window for nuclear fusion remains open a crack because, although much delayed and very costly, it promises almost unlimited emission-free energy without the near-eternal waste storage problem of nuclear fission.

Although they are novel and seductive, methane hydrates have a limited future, either in power plants or vehicles. As proficiency, efficiency, and scale bring down the cost of methane hydrates, they might become competitive in certain situations. However, time is not on their side. The window for any and all fossil fuels is closing.

Dimension 2

Weather and Climate

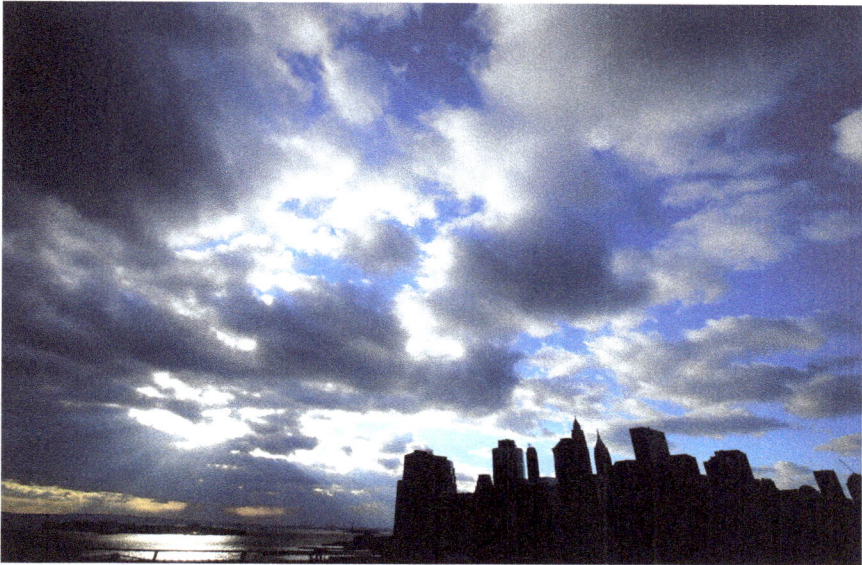

Downtown Manhattan and Governor's Island (New York City, 2011)

Weather and Climate

Weather has always snarled, roared, and raged from time to time. Lately, however, the elements have gotten more extreme. Storms are fiercer, with higher winds, bigger storm surges, more inches of rain, higher frequency, and longer duration. Heat waves are hotter and hang around past endurance. Droughts last for years, even decades. Once-in-a-hundred-year floods seem to happen with distressing regularity.

There is considerable confusion between the two terms — weather and climate. *Weather* is variable day to day — cold today, hot tomorrow, rainy the day after, sunny all weekend. *Climate*, on the other hand, is long term, its characteristics only calculable after 30 years or more. **Dimension 2: Weather and Climate** considers extreme weather events, many of which are increasing in frequency and duration due to human-caused climate change; the oceans, crucially important parts of the Earth's climate system; and the coastal regions that abut them.

Weather is inherently changeable. It can be sunny, cloudy, rainy, or snowy, sometimes all in the same day. As the well-known quotation (often attributed to Mark Twain) puts it, "If you don't like the weather in New England, just wait a few minutes." Yet the weather's nearly constant variation normally has limits. Those limits are not all that wide apart, and most of weather's variation occurs within those bounds. Occasionally, the range of a region's normal weather shifts due to large-scale variability patterns of high-altitude winds and ocean temperatures. Examples of this occur in northern Europe and southwestern America.

One such alteration was responsible for what has been called Europe's Little Ice Age, which lasted from early in the 14th century all the way into the 19th century. The Thames froze several times in London, stimulating frost fairs that flourished on the ice. Ice skating on Holland's canals became a favorite winter pastime. An example of another shift is the extended droughts that have visited the U.S. Southwest, some lasting for a century. The Anastasi and other indigenous cultures developed adaptations that permitted them to survive dry conditions for hundreds of years, but drought finally drove them to abandon their cliff dwellings for less arid lands. Scientists think that some parts of the Southwest are currently in the early stages of a drought that could last a century.

These extended perturbations to a region's weather are deviations from even longer-term norms. Since the end of the Little Ice Age in the mid-1800s, the northern hemisphere has experienced a long period of stable, warm weather. During this period, Europe and North America have experienced weather that has been fairly reliable as to temperatures and precipitation patterns. People came to know and depend on the yearly passing of the seasons. Farmers knew when to plant and when to harvest. Sure, there were late springs, hot summers, early falls, and harsh winters, but the variations rarely exceeded expectations of weather's normal variations.

Increasingly, weather is becoming less normal in many parts of the world. People perceive, and remark, that the weather is changing. Not only does spring come earlier than ever, but winters are milder and have lower snowfalls. And not only are the seasons changing but so is the degree of variation from the norm. As the level of greenhouse gases generated by burning fossil fuels rises in the atmosphere, more and more the weather acts abnormally, in a phenomenon known as extreme weather. Temperature records have been set repeatedly, only to be broken in short order.

Section 1 "Extreme Weather" examines how some weather phenomena that have always occurred — droughts, floods, storms, and heat waves — are being exacerbated by climate change, bringing greater perils. The year 2017 set records for hurricanes. On August 17, Hurricane Harvey, a Category 4 storm (with winds between 131–155 miles per hour), devastated Houston with heavy rains that overwhelmed the city's flood control system, resulting in whole neighborhoods inundated, 68 deaths, and $125 billion damage. In short order, Hurricane Irma, a Category 4 storm with 130 miles-per-hour winds, tore through the Florida Keys on September 10, causing much destruction before striking Marco Island. Hurricane Maria, a Category 5 with top winds of 174 miles per hour, then scored a direct hit on St. Croix, Dominica, and Puerto Rico from September 16 to October 2, knocking out the U.S. territory's power grid, leaving the populace without electricity and with minimal phone communication for months. All in all, Hurricane Maria contributed to more than 3,000 deaths.

Droughts have also been occurring more often, lasting longer, with greater severity. The Middle East experienced drought conditions from 1998 to 2012. California's longest drought began in 2011 and went through most of 2019, with drought recurring in 2020 and 2021.

Nicaragua and the Caribbean region endured a prolonged period of little rain from 2013 to 2016, with concatenating droughts in 2018 and 2019. Ethiopia's chronic drought conditions worsened in 2016, to the point that the UN and international relief organizations mounted major operations to provide aid to the starving people there, despite an ongoing civil war.

In July 2010, Pakistan was struck by devastating floods that killed some 2,000 people and left 11 million homeless. Causes of the floods include changed monsoon patterns and rainfall amounts, deforestation, and dams. Before Pakistan could fully recover, new severe floods swept the country again in August 2011, killing more than 200, destroying 670,000 homes, and affecting 5 million people.

Section 2 focuses on topics related to "Oceans and Coasts," which are inextricably linked to climate change. Spanning 70% of the Earth, oceans and seas continually interact with the atmosphere. Global heating is expanding the water in the oceans and is melting the land ice at both poles. Together, these phenomena contribute to causing sea level to rise, leading to aggravated flooding of coastal towns, cities, and wetlands.

The Atlantic Gulf Stream carries warm water from lower latitudes to North America and Europe. This warm water influences atmospheric flows that moderate not only the climate of both New England and the Mid-Atlantic states but also much of northern Europe. Weakening or failure of the Gulf Stream may have contributed to outbreaks of what was first called Justinian's Plague (541–750 AD) and later the Black Plague in the 13th, 14th, and 15th centuries. It has also been implicated in the Little Ice Age, generally dated from 1303 to 1860 in Europe, and the Great Famine that occurred there during that period from 1315 to 1317.

Turning to the southern hemisphere, the Pacific Ocean is subject to the El Niño Southern Oscillation (ENSO). Warm surface waters together with the trade winds that normally blow from East to West periodically reverse approximately every 3 to 8 years. When the reversal happens, the coast of South America experiences warm air and water currents known as El Niño events. El Niño tends to bring warm, wet weather to the west coasts of the Americas, and cool, dry weather to Australia and Asia. When ENSO swings to extremes in the East to West direction, a condition known as La Niña occurs. It has the reverse effect of El Niño, causing cool, dry weather in the Americas and the opposite in Asia. Between the El Niño and La Niña phases of ENSO are neutral periods when conditions in both Asia and the Americas are "normal."

Learning about weather and climate in Dimension 2 helps us to understand what climate change is all about.

Extreme Weather

Storm clouds (New York, 2008)

The Droughts and Floods of Climate Change

Climate change is contributing to recent occurrences of droughts and floods. Meteorologists have long warned that global warming will change precipitation patterns, leading to heavier storms, increased flooding, and prolonged and more intense droughts. According to the United States Environmental Protection Agency (EPA),[1] the rise in temperatures from global warming has caused an increase in the rate of evaporation, which means more precipitation in the form of rain in some regions of the world as well as less precipitation falling as snow.

On a radio interview on WBAI, leading climate scientists from Stanford University, Noah Diffenbaugh and Mark Jacobson, explained that a declining spring snowmelt dramatically impacts the water supply.[2] For example, the Folsom Dam in California had little inflow the last few years because the snowpack on the Sierra Nevada Mountains has been the skimpiest in a century. About a third of California's water storage depends on snowpack as a natural reservoir, unlike other regions that rely on concrete or manmade reservoirs to store melted snow.

As average temperatures have risen, so has rainfall. The Union of Concerned Scientists tells us this linkage happens because warmer air holds more moisture and when it meets cooler air, "the moisture condenses into tiny droplets that float in the air."[3] When the droplets become bigger, rain develops.

Downpours in Texas in May 2015 made for devastating floods that killed 15 people and left 12 missing. Texas A&M University[4] reported that record rainfall in many parts of Texas has significantly raised the amount of freshwater pouring into the Gulf of Mexico — to as high as 10 times the normal rate. The report quoted Steve DiMarco, professor of oceanography, saying that "rivers such as the Brazos, Trinity, Colorado, and others currently carry record amounts of water flowing southward to the Gulf." In 2007, the rivers also overflowed, carrying 10 to 20 times the normal seasonal rate of discharge into the Gulf, according to DiMarco. The floods created huge problems for marine life and commercial fishermen.

---

[1] https://www.epa.gov/, Environmental Protection Agency.
[2] https://www.wbai.org/, WBAI Radio 99.5FM NYC.
[3] https://www.ucsusa.org/, Union of Concerned Scientists.
[4] https://www.tamu.edu/, Texas A&M University.

February Hotter by Two Degrees

Scientists say that February of 2016 was hotter than usual, with average temperatures more than 1° Celsius (1.8° Fahrenheit) warmer than normal. February was the third month in a row with above normal temperatures, according to an analysis by NASA's Goddard Institute for Space Studies (GISS).[5]

The February air was warmer by 1.35° Celsius (2.43° Fahrenheit), topping January's record. The higher temperatures are alarming scientists because previous temperature spikes were usually by hundredths or tenths of degrees. NASA, which has been keeping a global temperature database that begins in 1880, says February's high temperatures were an extreme anomaly in their database. The Japanese Meteorological Agency[6] concurred with NASA in their global temperature analysis, saying the December–February period almost doubled the previous record warm anomaly for those three months, set during the 1997–1998 strong El Niño.

This hotter trend in 2015–2016 was attributed to a combination of a strong El Niño and human-caused global warming, caused largely by emitting greenhouse gases into the atmosphere. John Christy, Director of the Earth System Science Center at the University of Alabama in Huntsville believes that, "The record might have as much to do with an extraordinarily warm month in the Arctic as it does with warming caused by the El Niño."[7]

Climate change is usually assessed over multiple decades. Global average temperature in 2015, the hottest year on record since 1859, shattered the record set in 2014. Warming temperatures in 2016 broke that record for the third year in a row. The warmer waters off northern Australia raised concerns about massive coral loss. The spread of coral bleaching caused the Great Barrier Reef Marine Authority to issue a new alert, after divers discovered widespread loss of coral. Some areas in the Arctic experienced temperatures as much as 16° Celsius (about 29° Fahrenheit) warmer than average last month. Warmer weather was seen in

---

[5] http://data.giss.nasa.gov/gistemp/, NASA's Goddard Institute for Space Studies.

[6] http://ds.data.jma.go.jp/tcc/tcc/products/gwp/temp/feb_wld.html, Tokyo Climate Center, Japan Meteorological Agency.

[7] http://www.nsstc.uah.edu/nsstc/, The National Space Science and Technology Center, University of Alabama in Huntsville.

the northern hemisphere's higher latitudes. In 2016, large parts of Alaska and western and central Canada, as well as Eastern Europe, Scandinavia, and much of Russia, were at least 4° Celsius (roughly 7° Fahrenheit) above February averages, according to NASA/GISS.

Temperature data have been recorded routinely at land-based stations as far back as 1850. Climate scientists emphasize that whether a given month is a fraction of a degree warmer or cooler than a previous month is not as important as the long-term, overall trend. Disconcerting for many climatologists and experts is the dramatic rise of temperatures over recent decades.

Global Droughts and Climate Change

Globally, many areas are struggling with intense drought. One of the worst affected is Ethiopia, which has been dealing with drought for over 50 years, impacting over 10 million people. In 2016, the need for emergency aid increased significantly, since Ethiopian farmers had been unable to plant for two seasons, leaving the country without much of their usual food resources. International agencies working with the UN[8] aided the Ethiopian government, but the need for funds escalated dramatically. Requests for a $1.4 billion aid package saw less than half that amount pledged.[9] The International Rescue Committee (IRC) delivered clean water and helped to establish adequate sanitation. Children were at high risk because of malnutrition, and their days were taken up with the desperate search for water rather than going to school. The IRC fixed pump systems in a few villages to make groundwater more accessible.

In Nicaragua, a multi-year drought coupled with decades of deforestation nearly emptied the country's water sources including streams, rivers,

---

[8] http://www.africa.undp.org/content/rba/en/home/ourwork/climate-and-disaster-resilience/successstories/ethiopia-climate-change.html

[9] https://www.africanews.com/2016/02/01/14-billion-aid-needed-to-deal-with-ethiopia-drought/ "$1.4 billion aid needed to deal with Ethiopia drought," *Africanews*, February 1, 2016.

and lakes. By 2016, about 60% of Nicaragua's surface water sources had been lost and 50% of the country's aquifers had dried up or had become polluted. Also reported were the disappearance of at least 100 rivers and their tributaries and the contamination of Tiscapa and Nejapa Lakes near Managua, Lake Venecia on the west coast of Masaya, Lake Moyúa in northern Matagalpa, and the country's other large lake, Xolotlán, in Managua. Some of the country's major bodies of water had record low water levels, including the 420-mile-long Coco River, the longest in Central America, along the northern border with Honduras.

In the Middle East, climate change has contributed to what is believed to be the worst drought there nearly 1,000 years. A NASA study found that the Levant region — Israel, Palestine, Jordan, Lebanon, and Syria — was suffering from the worst drought in centuries. The researchers, who published their findings in the peer-reviewed *Journal of Geophysical Research*, studied the growth of tree rings in the region to determine drought variability across the Mediterranean Region. This data enabled them to document the region's historical climate. "Basically, we used a dataset of dry variability from the region that goes back, with reasonably good accuracy, to 1100 AD, and from that we were able to estimate that the recent drought in the Eastern part of the Mediterranean looks like it was the worst or driest drought anytime in the last 900 years," said Benjamin Cook, a climate scientist at NASA's Goddard Institute for Space Studies.[10,11] The Levant drought, which began in 1998 and lasted until 2012, became particularly severe between 2007 and 2010.

Drought in California has been almost continuous since the 2000s. In California, the four-year record-breaking drought between the fall of 2011 and the fall of 2015 was broken by much-needed rainfall in the winter of

---

[10] http://onlinelibrary.wiley.com/doi/10.1002/2015JD023929/full Cook, B. I., Anchukaitis, K. J., Touchan, R., Meko, D. M., and Cook, E. R., Spatiotemporal drought variability in the Mediterranean over the last 900 years, *J. Geophys. Res. Atmos.*, 121, 2060–2074, (2016); doi:10.1002/2015JD023929.

[11] https://www.nasa.gov/feature/goddard/2016/nasa-finds-drought-in-eastern-mediterranean-worst-of-past-900-years "NASA finds drought in Eastern Mediterranean worst of past 9 00 years," NASA, March 2, 2016.

2015–2016.[12] However, climatologists claimed that the drought was not over, and even though dry conditions have abated somewhat, the damage done by the drought to the state's water supply will be long-lasting. Long-term reserves in groundwater have been drained to the point that years, even decades, of wet weather would be needed to replenish them. Climatologists warn that the changing climate will require permanent changes in water-usage habits, but will Americans accept or resist such changes?

Record High U.S. Temperatures

What are the odds of there being 51 record high temperatures in the United States in November 2016 to only one record low?[13] Not very high. So we can safely assume that something besides pure chance is operating to skew temperatures towards record highs. That something is climate change.

In 2016, the average temperature for the whole year of the contiguous 48 states (not including Alaska and Hawaii) was 12.7° Celsius (54.9° Fahrenheit). According to NOAA,[14] that is 1.6° Celsius (2.9° Fahrenheit) above the 20th-century average and the second warmest year on record. The year 2016 was the 20th consecutive year that the annual average temperature was above the 20th-century average for this area.[15] An overall

---

[12] https://www.latimes.com/business/hiltzik/la-fi-hiltzik-drought-20160404-snap-html-story.html "No, California's drought isn't over. Here's why easing the drought rules would be a big mistake," Michael Hiltzik, *LA Times*, April 4, 2016.

[13] If you toss a coin, the odds are equal that it will land heads or tails. It is an even bet, 50–50 heads or tails. Say it lands heads. What are the odds that a second toss lands heads? Still 50–50. Say it lands heads again. A third toss? Still 50–50. However, if you ask a different question: What are the odds that two coin tosses both land heads? The odds are 1 in 4 — HH, HT, TH, TT. What are the odds of three tosses all landing heads? The odds are 1 in 8 — HHH, HHT, HTH, HTT, THH, THT, TTH, TTT. What about four tosses? 1 in 16. And five tosses? 1 in 32.

[14] https://www.ncdc.noaa.gov/sotc/national/201611, National Centers for Environmental Information, National Oceanic and Atmospheric Administration.

[15] https://www.scientificamerican.com/article/2016-was-the-hottest-year-on-record/ "2016 was the hottest year on record," Andrea Thompson, *Scientific American*, January 18, 2017.

warming trend in the U.S. has been occurring since 1895, with each decade since then experiencing an average temperature increase of about 0.1° Celsius (0.15° Fahrenheit) per decade.

Not only did high temperatures hit records in 2016, average low temperatures did not get as cold, with six states experiencing warmer than usual records as well.

The fall of 2016 was especially warm. According to NOAA, it was the warmest fall on record for the contiguous U.S. for the second straight year. Eight states experienced record warmth: Colorado, Iowa, Kansas, Michigan, Minnesota, New Mexico, Texas, and Wisconsin. Another 23 states had their second or third warmest fall on record.

November 2016 was an especially warm month in a record-warm fall season. The average temperature in November of that year for the contiguous states was 3.5° Celsius (6.3° Fahrenheit) above the average in the 20th century. States that were especially warm included Idaho, North Dakota, and Washington. NOAA reports that during November 2016 there were 40 times more record warm records than cold ones.

Why was 2016 so warm not only in the U.S., but across the world? A large El Niño — a climate pattern that brings unusually warm surface waters to the eastern tropical Pacific Ocean — played a role in raising global temperatures during both 2015 and 2016.[16] However, a substantial part of the warming in 2016 was due to heat trapped by greenhouse gases that humans have been emitting since the beginning of the Industrial Revolution, particularly carbon dioxide due to the burning of fossil fuels and the clearing of land for agriculture.

One silver lining associated with the record warm temperatures in the fall and winter was that demand for energy fell to the third-lowest value on record, due to reduced need for heating. However, does saving a few dollars on our heating bills justify heating the planet indefinitely? A reminder — high temperatures in the summer increase demand for air conditioning, so our energy bills would go up anyway.

---

[16] https://www.nationalgeographic.org/encyclopedia/el-nino/, El Nino, National Geographic.

Good evidence shows that the weather is getting warmer and the warming is caused primarily by humans. Unless we act to reduce the emissions of carbon dioxide and other so-called greenhouse gases caused mostly by burning fossil fuels and land clearing, the weather will only continue to get hotter.

What can we do about it? Drive less, walk or cycle more, lower the thermostat and water heater, put on a sweater. Run the dishwasher and washer/drier less often. Buy selectively at the supermarket and throw out less food (it takes energy to grow and get it to you). Switch off the lights when you leave a room. Turn in your gas guzzler car for a higher mile-per-gallon model or a hybrid or an electric vehicle. Do not fly if you can take a train or a bus. Recycle. Urge your congressperson, state representative, and local council member to support green climate policies. Ask your friends and neighbors to do the same. The personal matters, and so does policy.

Update: At the time, 2016 was the hottest year on record for the globe. This record was held uniquely by 2016 until 2020, when 2020 tied the 2016 record. Overall, the average temperature of the Earth has increased more than 1° Celsius (1.8° Fahrenheit) since the 1880s.[17]

Megastorm Harvey

Climate scientists and meteorologists alike know that Hurricane Harvey occurred in one of the most hurricane-prone regions of the world, at the peak of hurricane season in August 2017. This hurricane, however, had extraordinary destructive force. Over 120 centimeters (nearly 50 inches) of rainfall fell in the Houston area. To understand why, many weather specialists are now considering how much climate change and global warming may have contributed to this hurricane's intensity. The consensus is that climate change did not cause Hurricane Harvey, but it very probably made it worse.

---

[17] https://www.nasa.gov/press-release/2020-tied-for-warmest-year-on-record-nasa-analysis-shows "2020 tied for warmest year on record, NASA analysis shows," NASA, January 14, 2021; updated March 17, 2021.

In the wake of this colossal hurricane, opinions abound on the role that climate change played in the storm's massive force and devastation. For example, climatologist Michael Mann,[18] speaking with German Lopez of Vox.com,[19] stressed how climate change "exacerbated several characteristics of the storm in a way that greatly increased the risk of damage and loss of life." Mann prefaced his remarks by noting it was "unclear if global warming will lead to more storms like Harvey."

However, it was clear, Mann said, that climate change had previously led to a 15-centimeter (6-inch) increase in Houston's sea level, making the region more prone to dangerous flooding. In addition, Mann continued, "rising temperatures in the region add up to 1°C (1.8° Fahrenheit) to 1.5°C (2.7° Fahrenheit) higher than temperatures than a few decades ago." As Mann points out, this rising temperature has led to approximately 3% to 5% more moisture in the air and more risk of dangerous rainfall and flooding, as seen in Hurricane Harvey.

In addition to the rising sea level, warmer temperatures, and increased moisture in the air, Robinson Meyer in the *Atlantic* observes that the temperature of the "sea-surface waters near Texas rose to between 1.5°C and 4.0°C (2.7° and 7.2° Fahrenheit) above average.[20] These waters were some of the hottest spots of ocean surface in the world," enabling Hurricane Harvey to feed "off this unusual warmth" and "to progress from a tropical depression to a Category 4 hurricane in roughly 48 hours." Kevin Trenberth, a senior scientist at the U.S. National Center for Atmospheric Research, was quoted in the *Atlantic* saying, "The human contribution can be up to 30 percent or so of the total rainfall coming out of the storm … It may have been a strong storm, and it may have caused a lot of problems anyway — but [human-caused climate change] amplifies the damage considerably."

---

[18] http://michaelmann.net/, Michael Mann.

[19] https://www.vox.com/science-and-health/2017/8/28/16214268/houston-floods-harvey-global-warming "How global warming likely made Harvey much worse, explained by a climatologist," German Lopez, *Vox*, August 28, 2017.

[20] https://www.theatlantic.com/science/archive/2017/08/did-climate-change-intensify-hurricane-harvey/538158/, "Did Climate Change Intensify Hurricane Harvey?" Robinson Meyer, *Atlantic*, August 28, 2017.

In a series on Houston's flood risk on propublica.org, reporters looked not only at how climate change will bring more frequent and fierce rainstorms to cities like Houston but also at how growing development and urban sprawl in this loosely zoned city are creating greater flood risks. Houston's unchecked growth has seen builders ignoring strict regulations by paving over natural prairie land that once could absorb large amounts of rainwater. In a storm like Hurricane Harvey, this phenomenon played out by turning streets into waterways, which overflowed into the bayou network, drainage systems, and reservoirs. As Texas struggles with rebuilding after this mega-hurricane, Houston in February 2020 inaugurated a comprehensive resiliency plan to prepare for future hurricanes.[21]

Heat Projections Higher in Recent Climate Models

Global climate models (GCMs) have been used by climate scientists since the 1970s. They have been refined over the years as more data have accumulated and computers have become faster and more powerful. Yet, even the relatively simple models from the 1970s have proved to be quite accurate at predicting future climate. Most have predicted about a 3° Celsius (5.4° Fahrenheit) rise in global temperature by 2100 under "business-as-usual" greenhouse gas emissions conditions. Yet over the past year, the predictions of some of the new versions of the climate models are "running hot," predicting roughly a 5° Celsius (9° Fahrenheit) rise by the end of the century.[22]

One possible reason is a recently improved module in some GCMs to simulate the effects on global temperature of clouds and aerosols in the atmosphere. The question the climate modelers are scrambling to answer is whether the new module is wrong, overweighting cloud impact on temperature, or if it is accurate, which could mean that previous models

---

[21] https://projects.propublica.org/houston-cypress/, Boomtown, Flood Town, *Texas Tribune*.

[22] https://www.bloomberg.com/news/features/2020-02-03/climate-models-are-running-red-hot-and-scientists-don-t-know-why, "Climate Models Are Running Red Hot, and Scientists Don't Know Why," Eric Roston, *Bloomberg News*, February 3, 2020.

understated that impact, an unlikely outcome given the general accuracy of past models.

The predicted temperature spike may be unrelated to the refined cloud module. Some other sensitivity or combination of sensitivities may be responsible for the jump. Scientists have identified almost a dozen "tipping points," any one of which, if passed, would have cascading impacts on other parts of the hugely complex climate system. The observed anomalous spikes in the latest climate models could be telling us that one or more of those tipping points has been reached, with the anticipated calamitous implications for our climate future. Bad news for the human race and indeed all life on Earth if that is the case.

Texas Freeze and the Polar Vortex

In 2021, fluctuations to the polar vortex were responsible for the coldest February since 1989. According to the World Meteorological Organization (WMO), natural disruptions caused the polar vortex to weaken, wobble, and slip southward off the pole. The extreme weather event brought record-low temperatures to a large portion of central Canada, the United States, and even northern Mexico. Temperatures recorded during February fell as much as 14 to 28° Celsius (25° to 50° Fahrenheit) below average.[23]

Southern states in the U.S. were the most affected by the descending Arctic air, particularly in Texas and parts of Oklahoma and Arkansas. Other weather patterns interacted with the cold front to cause an unprecedented band of freezing air to linger over the South for most of early February.

Furthermore, severe winter storms during February brought heavy snowfall and treacherous icy conditions to large communities such as Houston, which rarely see snow at all. In Texas, the cold front caused damage by putting significant strain on the municipal power grid and

---

[23] https://www.nbcnews.com/science/environment/topsy-turvy-polar-vortex-brought-record-freeze-texas-rcna290, "How 'topsy turvy' polar vortex brought record freeze to Texas," Associated Press, NBC News, February 17, 2021.

water supply. Millions of Texans lost power during the storm and were unable to heat their homes or prevent their water pipes from bursting. At least 217 people were killed directly or indirectly by the severe cold, and property damages were estimated to be at least $195 billion.

Making matters worse, the state's power plants were not prepared to handle the sustained freezing conditions, which are rare in Texas. The power grid failures led to further shortages in basic food supplies and clean drinking water, as the winter storm interfered with supply chains and municipal plumbing. These shortages also made it harder for hospitals to care for thousands of patients suffering from hypothermia (subnormal body temperature) and other injuries caused by the storm. For those sticking it out at home, Texans were advised to boil their tap water to prevent risk of illness until wastewater treatment facilities could get back up and running.

Natural gas was hit the hardest of all the utilities in Texas because both the production and distribution pipelines froze and shut down. The problem only worsened as millions of freezing Texans turned up their heat. The abrupt spike in demand for natural gas for heating purposes led to shortages at power plants using natural gas to produce electricity. Some natural gas providers even took their power plants completely offline to avoid paying the high cost of fuel and to avoid long-term damage to their facilities. With coal, wind, and nuclear power plants also stressed with their own power shortages, the state's energy grid collapsed in a time of great need.

*****

In recent years, "polar vortex" has become a commonly used term during winter weather reports throughout North America. While many people would recognize this phrase or use it to describe colder-than-usual weather, its scientific meaning is complex, and the natural phenomenon's connection to climate change is still largely unknown.

In essence, the polar vortex is a band of westerly winds that accumulates between 10 and 30 miles above the North Pole every winter. (There is a similar polar vortex at the South Pole, but it has less effect on weather in the Northern Hemisphere.)[24] Parallelling the vortex at a lower altitude in the troposphere — five to nine miles above the surface — the polar jet stream functions to keep the large vortex of extremely cold air contained in the stratosphere.

Isolated from warmer latitudes by the polar jet stream below, the air inside the vortex grows progressively colder throughout the winter. During a normal winter when the polar vortex is stable, the polar jet stream is able to shift northward and firmly contain the vortex of cold air. In this scenario, the coldest polar air remains in the Arctic, resulting in milder weather in the mid-latitudes.

Conversely, in years when the polar vortex weakens or shifts, the polar jet stream can become wavier, lose strength, slow down, or sometimes reverse. These disruptions to the polar vortex's usual position are called sudden stratospheric warmings because they allow warm air from the south to be pulled into the Arctic, which can contribute to glacial melt. When this northerly movement of warm air happens, the displaced polar air is pushed south into the mid-latitudes, which can consequently cause the freezing temperatures and inclement winter weather the polar vortex is known for in America. This phenomenon naturally occurs every few years but is hard to predict or prepare for.

*****

The extreme cold in the Southern Plains during February 2021 also coincided with warmer-than-average temperatures in the Arctic. It follows to wonder if changes in the frequency and severity of polar vortex fluctuations are caused by or are contributing to global warming. Are severe winter weather outbreaks in the mid-latitudes becoming more likely?

Because disruptions of the polar vortex occur when the polar jet stream below it is altered, any forces — including changes to land and sea surface temperatures that impact the strength or location of the polar jet

---

[24] https://www.nytimes.com/2021/02/20/us/texas-winter-storm-explainer.html, "Texas Winter Storm: What to Know," Nicholas Bogel-Burroughs, Giulia McDonnell Nieto del Rio and Azi Paybarah, *New York Times*, February 20, 2021.

stream — can act to initiate a release of cold air from the Arctic. It appears that warming in the Arctic and diminishing sea ice are causing the polar jet stream to meander irregularly in at least two places, which makes it more likely to disrupt the polar vortex. So, it is plausible that even though there is an overall warming trend on Earth due to human-caused greenhouse gas emissions, we may see an increase in the severity of winter weather events in the mid-latitude regions of North America and Eurasia, due to interactions of the warming with the polar vortex.[25]

On the other hand, outbreaks of cold Arctic air are natural and have been a known part of our regular seasonal weather patterns in North America since the pattern was first recorded in the 1950s. Sometimes the polar vortex can be disrupted without significant impacts on surface weather in lower latitudes. For these reasons, blaming the polar vortex and human-caused climate change for colder winter weather in America might be seen as a bit of a stretch.

There are still many things climate scientists do not understand about the polar vortex, so it is hard to pinpoint climate change as the primary reason the polar vortex fluctuations are happening more regularly. Instead, scientists suggest the reason is likely more of a mix of random natural weather patterns and human-induced climate changes. While the polar vortex's exact links to climate change are yet to be ascertained, colder-than-normal winter weather events are not likely to negate the long-term warming trend from climate change.

In fact, the WMO reports that overall cold temperature records are becoming rarer as heat records and heat waves are becoming more common. This overall warming trend is anticipated to continue because worldwide greenhouse gas concentrations are on the rise. Even though February 2021 was a particularly cold month in North America, globally, the average carbon dioxide concentration in the same time period was 416.75 parts per million, up from 413.4 parts per million in February 2020.

---

[25] https://www.climate.gov/news-features/understanding-climate/understanding-arctic-polar-vortex, "Understanding the Arctic polar vortex," Rebecca Lindsey, *Climate.gov*, March 5, 2021.

That the southern U.S. experienced a relatively cold winter in 2021 should not be seen as proof that global warming is not happening. It is. We just do not have a very long record of data on polar vortex fluctuations to know what patterns are normal (or random) and what patterns are being caused or intensified by climate change. Some climate models predict that continued warming and sea ice melt will lead to a weakening of the polar vortex. Other models predict the opposite. This discrepancy is largely because the correlation between Arctic surface and sea temperatures and the atmospheric systems above are intricate and have only been studied for several decades.

With the data available at present, the effect of global warming on the polar vortex appears to be minimal compared to the natural variability of the phenomenon. Any future influence the polar vortex may have on winter weather is also predicted to be small in relation to the overall warming influence of greenhouse gas emissions. On average, our winters in North America are still warmer than in the past, and record-setting cold extremes have become far less likely.

*****

Even if extreme winter weather events caused by the Arctic polar vortex occur infrequently, their potential damage to communities in the southern U.S. and elsewhere around the world is significant and is worthy of further research and monitoring. The precise linkages between polar vortex fluctuations and climate change may still be unknown, but the more we can learn about this intricate relationship the better. Increasing our understanding of the polar vortex and how to predict its movements will extend the lead time both citizens and government officials have to prepare for the next big storm.

Oceans and Coasts

Portland Head Light (Maine, 2020)

If You Live on the Coast, You're Gonna Get Wet

A study published by the Union of Concerned Scientists[26] says that already 90 cities in the U.S., mostly in Louisiana and Maryland, are inundated, which is defined as a non-wetland area subject to flooding 26 or more times a year.[27] Within the next 15 to 20 years, that number will rise to 165 to 180 cities, mostly in New Jersey, North Carolina, and Louisiana. Within 40 years, 270 to 360 cities will face disruptive high tides and flooding that make conducting business as usual impossible. By the end of the century, using a moderate scenario for sea level rise, 490 cities will be inundated, more than 50 of which will have 100,000 people or more. That list includes Boston, Fort Lauderdale, Houston, San Francisco, Los Angeles, and four of New York City's five boroughs.

If the world can keep the global temperature rise to 1.5° to 2° Celsius (2.7° to 3.6° Fahrenheit) by 2100, between 200 and 380 coastal cities would be spared devastation. Cities will have to plan and act to prevent, adapt to, or retreat from the flooding, as will residents, businesses, and public institutions. Concerted government action is a must, such as setting policies for strategic relocation and providing resources to implement them.

Stay or Go? Pay or Pray? Rebuild or Relocate?

Hurricane Harvey flooded Houston, but altered temperature and rainfall patterns, as well as more severe storms, are likely to put America's entire real estate market in turmoil. Climate change is threatening property values throughout the United States. Millions of property owners in both coastal and low-lying areas will be faced with a difficult reassessment of risk and the costs and benefits of strategic relocation. Hurricane Katrina and Hurricane Sandy may have been shrugged off as anomalies, once-in-a-century "perfect storms," but Hurricane Harvey, the third recent

---

[26] https://www.ucsusa.org/resources/when-rising-seas-hit-home, "When Rising Seas Hit Home: Hard Choices Ahead for Hundreds of US Coastal Communities," Union of Concerned Scientists, July 5, 2017.

[27] http://i2.cdn.cnn.com/cnnnext/dam/assets/170711201543-weather-how-sea-level-rise-causes-chronic-inundation-exlarge-169.jpg, infographic on sea level rise.

megastorm, establishes a pattern. It sets the new normal. It forces a re-evaluation.

For sure, insurance companies will alter their risk-assessment models to incorporate the new reality. Then they will revise their rates, the premiums they charge property owners for providing flood insurance. Sharply higher rates will force many property owners to face difficult choices: stay or go, pay or pray, rebuild or relocate? Towns and cities will face similar decisions. How much adaptation can we afford? How much can we raise taxes? Can this neighborhood be saved or must its citizens be relocated?[28]

Low-lying areas will deteriorate and would not be rebuilt quickly, if ever, as areas on higher land gentrify and gain in value. Legislatures at state and federal levels will have to reform flood insurance programs, tightening coverage requirements and enforcement mechanisms. People's outcries and anguish will be real. As usual, the poor will pay the most, but so widespread will be the effects of real estate's new reality check that politicians will have to pass reforms.[29]

A vast reshuffling of property values, living patterns, and political power will flow from climate change, in a slow-moving wave that will take years to ripple through the U.S. and other countries. After Hurricane Harvey, the country should be more aware of climate change and more attuned to issues surrounding it. The hope is that real estate brokers will become proactive in regard to protecting their clients from increasing risks of major storms, sea level rise, and coastal flooding.

The Sea Level Rise and Coal Connection

Significant sea level rise (SLR) is inevitable. Whatever scenario of climate change plays out, we will have to take steps to adapt to more coastal

---

[28] https://www.washingtonpost.com/opinions/after-harvey-flood-insurance-needs-reform/, op-ed, *Washington Post*, August 30, 2017.

[29] https://www.theguardian.com/environment/2017/aug/29/hurricane-harvey-climate-change-real-estate-florida?CMP=Share_AndroidApp_Gmail, "How climate change could turn US real estate prices upside down," Richard Luscombe, *Guardian*, August 29, 2017.

flooding and higher storm surges. Some projections, including those that incorporate the potential impact of melting in Antarctica, suggest oceans may rise by 1.3 meters (4 feet 3 inches) by 2100, even if the world keeps within the Paris Agreement target of limiting temperature rise to well below 2° Celsius (3.6° Fahrenheit).[30] This level is 50% above the 2015 projections of the Intergovernmental Panel on Climate Change (IPCC).[31]

Even more concerning, research done at the University of Melbourne has projected that global sea levels may rise 1.3 meters (4 feet 3 inches) by 2050, unless coal power ends.[32] The research looked at all the factors contributing to SLR and suggests that if coal power generation falls to near zero by 2050 instead of 2100, carbon emissions would decrease enough to keep global temperature rise below 1.9° Celsius (3.4° Fahrenheit). Less use of coal would also mean that drastic Antarctic melting would not be triggered, thus halving the estimated sea level rise.

Even if you do not personally live near a coast, millions of people do. Surely avoiding over 1.2 meters (4 feet) of repeated urban flooding should be enough incentive to start advocating for a rapid transition to low-carbon energy. Coastal and tidal municipalities need to take all feasible steps to prevent or minimize as much disruptive flooding as possible.

Rapid Warming in Utqiagvik, Alaska

A computer algorithm at a weather station in the northern Alaskan town of Utqiagvik (formerly Barrow) on the Arctic Ocean interpreted rising temperature data as an anomaly in 2017. In reality, it was climate change, occurring in a rapid, dramatic way that the programmers had not anticipated.

---

[30] http://iopscience.iop.org/article/ Nauels, A., Rogelj, J., Schleussner, C.-F., *et al.*, Linking sea level rise and socioeconomic indicators under the Shared Socioeconomic Pathways, *Environmental Research Letters*, 12 (11), (2017).

[31] https://www.ipcc.ch/report/ar5/, AR5 Synthesis Report: Climate Change 2014.

[32] https://www.theguardian.com/environment/2017/oct/26/sea-levels-to-rise-13m-unless-coal-power-ends-by-2050-report-says?CMP=share_btn_link, Michael Slezak, *Guardian*, October 26, 2017.

The National Center for Environmental Information (NCEI)[33] has been collecting weather data since the 1920s. However, equipment breaks down over time, so NCEI employed an algorithm to flag anomalous readings and omit outliers from its reports. Recently NCEI noticed that data from Utqiagvik was missing for over a year. Had their instruments broken down? Upon investigation, NCEI scientists discovered that the temperatures had been recorded, but they were so high that they were not reported.[34]

Since 1979 when measurements using the algorithm began, average January to September temperatures in Barrow went up 1.1° Celsius (1.9° Fahrenheit), at a rate about twice that in the rest of the U.S. However, in October to December 2017, the temperature gain was 4.3 degree sign, 3.8 degree sign, and 2.6° Celsius (7.8°, 6.9°, and 4.7° Fahrenheit), so high the algorithm kicked in and withheld the data. Turns out the equipment was not broken — the climate was. As the weather warmed up, more ice melted, which led to higher temperatures and still more ice melting, a cycle that led the *Smithsonian Magazine* to call Utqiagvik "ground zero" for climate change.[35]

Time to Sell Your Beachfront Property?

Satellite data over the last 25 years show an alarming rate of sea level rise. Scientists using satellites, and carefully correcting for such atmospheric anomalies as the eruption of Mt. Pinatubo, calibration errors, and the fact that the sea floor is sinking, have determined that sea level is indeed rising and is rising faster and faster each year.[36] The effects of rising sea levels

[33] https://www.ncei.noaa.gov/, National Center for Environmental Information (NCEI).

[34] https://www.huffpost.com/entry/arctic-temperatures-rising-fast-reported-false_n_5a316 487e4b07ff75affaa1f?dl%2C=, "Artic temperatures are rising so fast computers don't believe they're real," Ryan Grenoble, *Huffpost*, December 13, 2017.

[35] https://www.smithsonianmag.com/smart-news/goodbye-barrow-alaska-hello-utqiag-vik-180961273/, "Goodbye Barrow, Alaska. Hello, Utqiagvik," Jason Daley, *Smithsonian Magazine*, December 2, 2016.

[36] https://www.inverse.com/article/40005-geologists-sea-level-rise-barystatic-geocentric-seafloor-sinking, "Ocean levels aren't just rising, sea floor levels are sinking, too," Peter Hess, *Inverse*, January 9, 2019.

will be consequential for many coastal cities around the world. And, the rate of sea level rise is expected to increase as the climate gets warmer because melting of ice at both poles may speed up and because water expands as it warms. For coastal inhabitants, it is time to start thinking about strategic relocation.

Stolen Beaches to Pave Paradise

There is a worldwide shortage of sand. Why is sand in short supply? Thieves are stealing it, not to create other beaches, but as a raw material for the manufacture of concrete and aggregate.[37] Joni Mitchell was prescient — we have "paved paradise and put up a parking lot,"[38] as well as many high-rise condos and office towers. Ecosystems are being disrupted and demolished. Wildlife is being hounded, forced to adapt, move, or die. More than 50% of humans now live in cities; projections are that by 2100 that number will reach 85%.

Nobody quite knows what the implications of this growing urbanization are for humans, wildlife, or the climate. We are rushing headlong into the unknown. We can hope for the best, but we really ought to be planning more for the worst.[39] Humans are the most adaptable of creatures, but it behooves us to proceed with caution, lest we face the same choices as other wildlife: adapt, move, or die.

Why is the loss of sand important for climate change? At first, it is hard to see how any single change or factor can affect the climate. However, such factors as wide-scale environmental destruction, China's rapid urbanization, expanded transportation infrastructure, and rising sea levels do indeed impact the climate, both separately and in a dynamic that

---

[37] https://science.howstuffworks.com/environmental/conservation/issues/sand-is-such-high-demand-people-are-stealing-tons-it.htm "Sand is in such high demand, people are stealing tons of it," Dave Roos, *howstuffworks*.

[38] Mitchell, Joni. (1970). "Big Yellow Taxi." On *Ladies of the Canyon* [record]. Burbank, CA: Reprise.

[39] http://www.sandstories.org/, Sand Stories.

compounds the effects that each has individually.[40] Resources that were once thought infinite and inexhaustible — like land, clean air, oceans, forests and sand — are revealing themselves to be finite and exhaustible as this Anthropocene Era (the time of significant human impact on Earth) advances.

How Changing Currents Can Affect the Climate

Like most of nature, the Atlantic Meridional Overturning Circulation (AMOC) is complicated. To simplify, it is a multi-level ocean current that carries warm tropical water north to the Arctic, and cold Arctic waters south to the tropics. It moderates the climate of North America's east coast and Europe's northern countries. The air above the warm northward currents carry rainfall needed for crops to flourish on both continents. Failure or weakening of the AMOC has been suggested as a possible precipitating cause of the ice ages that have several times in the distant past buried northern parts of both Europe and North America under deep glaciers.

Scientists are learning more about it, thanks to satellites, vessel-borne measuring instruments, and buoys, but they do not have a long historical record. They have determined that the AMOC is slowing down, sparking worry that climate change is to blame and raising concerns that it might collapse and usher in a new ice age. Now, however, Hannah Hickey reports in her article about the AMOC that "Climate change isn't to blame for slower Atlantic circulation."[41] Scientists X. Chen and K. Tung, from China and the University of Washington, have found evidence that the climate is not the main reason for the current's slowdown.[42] Rather, they

---

[40] https://www.theguardian.com/global/2018/jul/01/riddle-of-the-sands-the-truth-behind-stolen-beaches-and-dredged-islands?CMP=Share_AndroidApp_Gmail, Neil Tweedie, *Guardian*, July 1, 2018.

[41] https://www.futurity.org/atlantic-ocean-circulation-climate-change-1816482/ "Climate change isn't to blame for slower Atlantic circulation," Hannah Hickey, *Futurity*, July 19, 2018.

[42] https://www.nature.com/articles/s41586-018-0320-y, Chen, X., Tung, K. Global surface warming enhanced by weak Atlantic overturning circulation. *Nature* 559, 387–391 (2018).

now believe, the AMOC is probably a natural periodic cycle that lasts 60 to 70 years. If the observed slowing is natural, it will eventually reverse.

Thus, we probably do not have to worry about a coming ice age. But a slower AMOC will mean that less warm salty water will be stored deep in the ocean, so the water at or near the surface will be warmer. These changes will be bad news for corals and other marine life. Tropical fish will seek cooler waters to the North, encroaching on Arctic species. Our climate will continue to warm, even faster than it has been over recent decades. Warmer surface water also suggests that hurricanes and tropical storms, which feed off of warm water, will get stronger and more frequent.

Board up and batten down, Florida and the Gulf Coast.

Dimension 3

Consequences
for Nature and People

October in New England (Maine, 2020)

Consequences for Nature and People

Climate change affects virtually everything on the planet. **Dimension 3: Consequences for Nature and People** focuses on the impacts of climate change on a panorama of systems and sectors. Section 1 "Deforestation, Fires, and Species Extinction" looks at natural area destruction in tropical regions, widespread burns in California and Australia, and biodiversity losses across the world. Section 2 "Famine and Food Security" considers how the changing climate is affecting food and agriculture and how soils can store carbon, thereby reducing greenhouse gases in the atmosphere. Heat waves and droughts contribute to crop failures that cost farmers their livelihoods, raise the price of food for consumers, and cause hunger among the poorest producers. Section 3 "Cities and Power Outages" covers climate change impacts in urban areas, including massive electricity failures, and how cities can develop resilience.

In Section 4, "Migration, Conflict, and Population," discussion centers around the overlapping stresses of climate extremes, war, and demographic dynamics. Many people around the world are experiencing displacement due to climate extremes. Conflicts can exacerbate such impacts due to disruptions in food supplies. For example, Ukraine has long been the breadbasket of the world, feeding millions with its exports of wheat and barley, but now the fields are not being planted and ports are blocked. The ripple effect on the global food supply is projected to be extreme — both for the poor who depend on aid but also for everyone, who might not have realized how much of their food comes from Ukraine. Droughts are making the situation even more dire as they occur in Africa and elsewhere.

Wildfires are a natural part of a forest's self-maintenance, but rising temperatures, longer and hotter droughts, and less snowmelt (which provides vegetation with needed moisture) mean that wildfires start earlier, burn more intensely, cover a wider area, and cause more destruction. Heat waves often feed droughts, which create conditions conducive to wildfires, whether the fuel is brush or forest. Given the intrusion of humans into once uninhabited areas, that destruction entails not just devastation of vegetation and wildlife, but also the loss of homes and human lives.

Some rain is normal, and farmers have come to expect rain in the amounts and at the times needed by the crops they traditionally choose to plant. They know that those varieties over many planting seasons have become adapted to those conditions. However, when the higher temperatures are significant and prolonged, as they often are when caused by elevated levels of heat-trapping chemicals in the air, the increased heat affects both the timing and the amount of rainfall that falls on a given field. These changes can lead to reductions in yield or even to crop failure.

Smallholder farmers can usually withstand one such failure, but a second failure can drive farmers away from their land to seek better living conditions, either relocating within their own country or emigrating to another. Within their own country, they may sell their labor to other farmers or move to the city in hopes of finding employment. To emigrate, in the hopes of reaching a land with better life chances, also entails great risk.

Climate change affects people, both as individuals and as communities. Many of the impacts of climate change are variations from the normal weather that people have learned over many centuries to accommodate to and prosper under. Broken temperature records now occur most often on the high side; few new record-cold temperatures have been experienced. The extreme hot temperatures tend now to stay for longer, causing heat waves that have proved fatal, especially to thousands of the most vulnerable human beings — infants, the poor, the elderly, and the ill.

The novel weather brought by climate change is almost by definition disruptive. Too much rain can be as harmful as too little. Torrential rains at the wrong time can destroy a crop or prevent it from being planted in a timely manner. Floods can devastate entire villages, forcing their inhabitants to leave, abandoning homes, fields, and livelihoods. People forced to flee their traditional lands acquire a new designation — climate refugees.

While climate change is not the only causative factor, several million people have been driven by drought and internal strife to emigrate from Sub-Saharan Africa to Europe in the past decade. Many smallholder farmers from the Northern tier of Central American countries were pushed from their farms by changed weather, societal breakdown, and civil war. In local cities, all too often they found only predatory gangs, so they attempt to better their lives by relocating to the U.S.

We have already seen several cases of societies being overwhelmed by climate refugees. In 2016 and 2017, millions of refugees from Sub-Saharan Africa tried desperately to immigrate to Europe. Initially, many countries accepted them, but as their numbers grew from a trickle into a flood, European countries increasingly withdrew their welcome mat, imposing limits and discriminatory regulations.

The scale of the immigrant wave — their sheer numbers — aroused fears of cultural dilution, economic effects on wages from additional workers entering the labor force, and increased taxes to pay for expanded social services. These cultural implications of immigration had — and have — political ramifications. Right-wing nationalistic parties swelled, weakening traditional parties, sometimes to the extent of shifting balances of power.

When Hurricane Maria devastated Puerto Rico in September 2017, islanders were left without electricity, communications, housing, and health care. Many opted to head for southern Florida, where a significant Spanish-speaking Hispanic community lives, primarily from nearby Cuba. Over the years, Miami has received thousands of political refugees from Cuba. This part of Florida has become a magnet for other Latin Americans, so that some Puerto Ricans already had family members living in the area.

Currently, the largest number of climate refugees is from the northern tier of Central American countries. These nations — including Guatemala, Honduras, and El Salvador — are experiencing both a long-term drought and climate change-related modifications to temperature and rainfall patterns — conditions that have made it impossible for many smallholder farmers to feed their families. Out of desperation, many have sought better life chances in the U.S.

Most unfortunately for them, the U.S. enacted policies that has made it more difficult for these refugees to enter the country, leaving some imprisoned in border camps and others left to fend for themselves in Latin American countries not much better off than the homelands they fled. As many policy makers and average citizens fail to realize, changing climate is a key factor that is driving so many people from their homes.

In short, **Dimension 3: Consequences for Nature and People** explores the multi-faceted impacts of climate change on the world we all depend on.

Deforestation, Fires, and Species Extinction

Native forest and cultivated fields in the village of Izotalillo (El Salvador, 2009)

Deforestation: There is Still No Planet B

The enormous problem of deforestation contributes to climate change in two main ways. When trees rot or are burned, they release their stored carbon back into the air via carbon dioxide (CO_2). Further, we lose the capacity of trees to remove CO_2 from the air and store it in their roots, leaves, and wood.

Forests are being cut down on a massive scale and at an accelerating rate. The world lost 47% more forested land in 2015 than it did 16 years before that.[1] In 2015, about 49 million acres of forest disappeared worldwide, an area equivalent to the size of Nebraska.[2]

Seventy percent of the trees in Indonesia are being cleared to grow palm oil trees. Brazil is cutting down vast areas of rainforest for logging and agriculture. While exact figures are hard to determine, in 2020, approximately 80,000 acres of forest were lost each day globally, especially in Brazil, Indonesia, and the Democratic Republic of Congo. Many more acres are degraded by development or other damage.

How much does deforestation contribute to global warming? Scientists are not sure, but they are working on the answer to this question. For now, all they will say is that if we collectively do not slow the rate of deforestation, we have little chance of meeting the Paris Agreement target of limiting the global temperature rise to well below 2° Celsius (3.6° Fahrenheit).

So, add slowing or stopping deforestation to humanity's list of "must do" tasks. Others include burning less fossil fuel and capturing and storing carbon in soils. (See Conclusion chapter for a full planetary to-do list and a compendium of climate change actions.)

There is still no Planet B.

---

[1] https://www.globalforestwatch.org/, Global Forest Watch.

[2] http://www.salon.com/2017/07/30/a-nebraska-sized-area-of-forest-disappeared-in-2015_partner/; and https://www.climatecentral.org/news/mississippi-sized-area-of-forest-disappeared-21641 "A Nebraska-sized area of forest disappeared in 2015," Bobby Magill, Climate Central July 30, 2017.

Forests Laying Down on the Carbon Job

Climate change is not all caused by humans, at least not directly. Turns out that scientists are witnessing the inability of forests to regulate and absorb carbon dioxide. Tropical forests have been seen as a stabilizer to the growing carbon and greenhouse gas emissions that are affecting global climate. In 2017, NASA reported that the satellite known as the Orbiting Carbon Observatory-2 (OCO-2) detected the largest annual rise of CO_2 within the atmosphere in at least 2,000 years. The satellite images seemed to confirm that the natural carbon removal service provided by tropical forests is dissipating and the forests have become net emitters of CO_2, thereby contributing to climate change.[3, 4]

Scientists think this reversal can be attributed to the very strong El Niño event the Earth experienced in 2015–2016. The droughts that accompanied the El Niño caused myriad trees to die. Not only did they fail to extract CO_2 from the air but they rotted and released CO_2 into the air.

Humans did not help by widespread clearing of forests for agriculture and timber, especially in Brazil and Indonesia. Deforestation is thus compounding the ability of forests to soak up CO_2.

Unfortunately, more atmospheric CO_2 means more climate warming and a higher likelihood of more strong El Niños, which in turn means that we will have to reduce our own carbon emissions ever further and faster to preserve a livable world for our children and theirs, and theirs.

Amazon Fires Threaten Doomsday

A doomsday tipping point may be approaching in the Amazon in which rising temperatures dry out the rainforest, making it increasingly

---

[3] https://science.sciencemag.org/content/358/6360/230, Baccini, A., Walker, W., Carvalho, L., *et al.*, Tropical forests are a net carbon source based on aboveground measurements of gain and loss, *Science* 358 (6360), 230–234 (2017), DOI: 10.1126/science.aam5962.

[4] https://qz.com/1102841/tropical-forests-caused-the-highest-co2-levels-in-2000-years-says-nasas-orbiting-carbon-observatory/, "Tropical forests caused the biggest atmospheric CO_2 increase in 2,000 years," Elijah Wolfson, *Quartz*, October 15, 2017.

susceptible to the spread of fires. Most of the fires are set by loggers, farmers, and ranchers clearing the forest, actions encouraged by Brazil's president, Jair Bolsonaro.[5]

The Amazon currently absorbs 5% of total global carbon emissions. As rising temperatures dry it out, the Amazon's capacity to absorb carbon decreases, and the rainforest becomes more prone to severe and widespread fires. In 2019, 73,000 fires were burning, twice the number of fires burning in 2018.[6] As the Amazon forest burns, stored carbon is released into the atmosphere. The carbon balance — absorbed versus released — could tip negative by 2050. If the entire forest of Amazonia burns, it would release 140 billion tons of carbon dioxide (CO_2) into the air, an event that would accelerate global warming, pushing it well past the Paris Agreement ceiling of 2° Celsius (3.6° Fahrenheit).

The causes and effects of this doomsday scenario are well known. If it comes to pass, the result would be catastrophic, especially in light of the dramatic acceleration in the rate of global heating and the vast increase in the number of fires. Especially in 2019 and again in 2020, the number and extent of the fires assaulting the Amazon rainforest have set new records.[7,8]

Since Bolsonaro's term does not end until January 1, 2023, his policy encouraging commercialization of the Amazon rainforest will continue at least until then. That continuation means the tipping point will undoubtedly occur well before 2050. While the date when the Amazon rainforest reaches a tipping point where it can no longer support the biodiversity its

---

[5] https://www.usatoday.com/story/news/world/2019/08/22/amazon-fire-ngos-brazil-president-jair-bolsonaro-suggests-cause/2082066001/, "Why is the Amazon rainforest on fire," Elizabeth Lawrence, *USA Today*, August 22, 2019.

[6] https://www.insider.com/amazon-fires-may-help-dieback-emit-carbon-hurry-climate-change-2019-8, "Fires in the Amazon could be part of a doomsday scenario that sees the rainforest spewing carbon into the atmosphere and speeding up climate change even more," Sinéad Baker, *Insider*, August 22, 2019; https://www.insider.com/author/sinead-baker.

[7] https://www.bbc.com/news/world-latin-america-54779877 "Amazon fires: Year-on-year numbers doubled in October," *BBC News*, November 2, 2020.

[8] https://www.bbc.com/news/topics/c4em684z247t/amazon-fires, various articles on "Amazon fires," *BBC News*.

intact ecosystem enables is unknown, Brazil's president has made his intentions clear that he would like to see it all "industrialized" as soon as possible.

We can and must publicize the alarming events happening in Brazil so as to convince Bolsonaro that the Amazon is more valuable to Brazil and the world as a rainforest rather than as a region of commercial farms and ranches. We can and should contribute to the charities working to slow the destruction, such as the Rainforest Trust and Amazon Conservation. The goal is the forest's continued sequestration of vital quantities of carbon — for its biodiverse wildlife and the indigenous peoples who live in and sustain the rainforest. And the larger goal is a planetary environment still conducive to the existence of the human race and all other species.

Flames, Smoke, and Embers: Australia Burns

Bushfire! An extreme climate-impacted event. An apocalypse of raging red flames burning out-of-control and a dark inferno of wind-whipped smoke, indiscriminately scorching woodlands and structures, inciting primal fear, panic, trauma, and blackening all in its path.

Australian bushfires[9] are generally described as uncontrolled wildland combustion burning in a grass, scrub, or forested area.[10] Such conflagrations vary as seasonal weather elements — wind, temperature, humidity, and rainfall — affect bushfire severity and spread. Most commonly, these fires are sparked by lightning strikes on dry vegetation, but sometimes from human influences — mostly accidental, such as a stray spark from farm machinery or spontaneous combustion of organic matter such as brush or litter. Once a field fire starts, embers blown by wind "spot," that is ignite, new areas and furiously spread blazes. Bushfires are the result of extreme weather conditions acting on dessicated vegetation to fuel so-called "super fires."

Australia battles bushfires every year, but the December 2019–January 2020 fire season was decisively one of the worst. It started much earlier than usual, with more fires getting very big, very quickly. Starting in September

---

[9] Bush is an Australian term for natural undeveloped areas.

[10] https://en.wikipedia.org/wiki/Bushfires_in_Australia, Bushfires in Australia, Wikipedia.

2019 in New South Wales (NSW) (subsequently occurring in six Australian states), massive bushfires formed that became "coupled" with the atmosphere, generating their own lightning and gusty, violent, unpredictable winds, fomenting thunderstorms and fresh voracious fires.[11] Fueled by a combination of extreme heat, unusually long drought, and strong winds, numerous wildfires swiftly escalated and raged out of control.

With thousands of property-owners and holiday-makers seriously threatened, first New South Wales, then the Federal Government declared a State of Emergency in November 2020. Defense Force personnel and equipment could then assist Emergency Service Management, Fire Authorities, and professional, volunteer, and international firefighting crews as they battled numerous wildfires in hellish conditions.

Bronwyn Adcock in *The Monthly* of February 2020, vividly conveys what she titles "Living Hell," the on-going horror of eastern New South Wales' bushfire siege. She experienced the prolonged terror and has eloquently written of the communities that were affected as they responded to threat, dislocation, loss, and grief.

> The enormity of roiling greyness, searing heat, capricious winds, scorched stench, crackling tree-top foliage, roaring house-high flames! Roads blocked, helicopters buzzing overhead like giant dragon flies, fire trucks and heavy vehicles "in battle"; the desperate worry, utter helplessness of people in the path of Nature's fury. Where will it strike next? Radio news up-to-date or delayed? Not knowing! Life and death decisions to go or stay! Loss of [family] photos, food, furniture, farm and forest; the blackened ruins of homes cordoned off for safety.[12]

As voracious blazes intensified in the days leading up to New Year's Eve, thousands of people were forced to evacuate or seek shelter on the New South Wales and Victoria beaches.

---

[11] https://www.theguardian.com/environment/2020/jan/13/explainer-what-are-the-underlying-causes-of-australias-shocking-bushfire-season "Explainer: what are the underlying causes of Australia's shocking bushfire season?" Graham Readfern, *The Guardian*, January 12, 2020.

[12] https://www.themonthly.com.au/issue/2020/february/1580475600/bronwyn-adcock/living-hell#mtr, "Living hell," Bronwyn Adcock, *The Monthly*, February 2020.

Massive evacuations played out in the hardest-hit regions of the south-east due to capricious winds pushing flames through towns. Accompanying smoke became another disaster. On January 1, Canberra recorded its worst pollution ever, with an air quality index many times higher than "hazardous." Smoke in Sydney even crept into birthing rooms, stopped MRI machines, and triggered respiratory distress.

In spite of 3,000 Australian Army Reservists bringing their own planes and equipment, there was not much more that firefighters could do until there was enough rainfall to stop the blazes or the fires ran out of fuel and burned themselves out. The fires in New South Wales were finally declared "contained" on February 13, 2020. The relief came after torrential rains marked the wettest week in the region for three decades.

*****

On February 4, 2020, Prime Minister Scott Morrison confirmed in a speech to Parliament[13] that 33 people had died, including nine firemen, three of whom were Americans, as a direct result of the 2019–2020 bushfires.[14] Smoke from the massive bushfires was linked to more than 445 deaths, a government inquiry subsequently found. In comparison, Australia's worst fire fatalities occurred on Victoria's 2009 "Black Sunday," when 173 people died and another 414 were injured. A 2020 Royal Commission thoroughly investigated the Bushfire Crisis, making numerous recommendations on improving preparedness and response procedures, including the provision of localized climate change projections as part of the national disaster risk reduction program.[15]

On February 12, 2020, the New South Wales Rural Fire Service (RFS) reported that 2,439 homes had been destroyed. And on February 28, the

---

[13] https://www.aph.gov.au/About_Parliament About Parliament, Parliament of Australia.

[14] https://www.aph.gov.au/About_Parliament/Parliamentary_Departments/ Parliamentary_Library/pubs/rp/rp1920/Quick_Guides/AustralianBushfires, "2019–20 Australian bushfires — frequently asked questions: a quick guide," Lisa Richards, Nigel Brew and Lizzie Smith, Parliament of Australia: Department of Parliamentary Services.

[15] https://naturaldisaster.royalcommission.gov.au/, The Royal Commission into National Natural Disaster Arrangements Report.

Australasian Fire and Emergency Service Authorities Council (AFAC) stated a combined total of 3,094 houses had been lost across New South Wales, Victoria, Queensland, the Australian Capital Territory (ACT), Western Australia, and South Australia.

The AFAC Report also recorded that over 17 million hectares had been burned across New South Wales, Victoria, Queensland, ACT, Western Australia, and South Australia. Between July 2019 and February 13, 2020, the NSW Rural Fire Service reported that 11,264 bush or grass fires burnt 5.4 million hectares (13.34 million acres). More than 80% of the World Heritage-listed Greater Blue Mountains Area and 54% of NSW components of the Gondwana Rainforests in Australia World Heritage property were affected by wildfire. However, some ecosystems, like the Australian eucalyptus forests, have a unique relationship to fire. Some trees actually depend on fire to release their seeds and will regenerate.

Australian wildfires have been declared among the "worst wildlife disasters in modern history." Estimates indicate more than one billion animals, birds, fish, and reptiles were killed in the bushfires, with more than 800 million of those in New South Wales. Sydney Science Professor Chris Dickman's January 8, 2020, calculations were deliberately conservative, with the University of Sydney's report stating that "the true mortality is therefore likely to be substantially higher than those estimated."[16] Around 25,000 koalas were feared dead on Kangaroo Island, South Australia alone.[17]

Fire management specialists and meteorologists assess the risk of bushfires using a forest fire danger index, which is a combined measure of temperature, humidity, wind speed, and the dryness, but, significantly, not the amount of ground fuel. Ongoing drought, surface soil moisture, wind speed, relative humidity, heat waves, dead and live fuel moisture, and land cover with certain vegetation (particularly native eucalyptus and

---

[16] https://www.aph.gov.au/About_Parliament/Parliamentary_Departments/Parliamentary_Library/pubs/rp/rp1920/Quick_Guides/AustralianBushfires, "2019–20 Australian bushfires — frequently asked questions: a quick guide," Lisa Richards, Nigel Brew and Lizzie Smith, Parliament of Australia: Department of Parliamentary Services.

[17] https://www.theverge.com/2020/1/3/21048891/australia-wildfires-koalas-climate-change-bushfires-deaths-animals-damage, "What you need to know about the Australia bushfires," Justine Calma, *The Verge*, February 13, 2020.

grazing land) all can be contributing primary causes to a fast-spreading bushfire phenomenon.

*****

While firestorms are not new to Australia, extreme heat and drought, which are two manifestations of climate change, create copious tinder to fuel them. Australia's weather is typically hot and dry, similar to conditions in California or the Mediterranean.

Rising levels of greenhouse gases in the atmosphere change the Earth's radiation balance, allowing less heat to escape. Australia suffered from its hottest day on record on December 18, 2019, reaching a national average temperature of 41.9° Celsius (107.4° Fahrenheit), ensuring Australia's hottest December on record. The year 2019 was the country's hottest and driest year on record (with average temperatures 1.52° Celsius (2.7° Fahrenheit) above the 1961–1990 mean and 40% lower rainfall than averaged records since 1900). The country's second hottest year was 2013, followed by 2005 and 2020 all since the year 2000.

Furthermore, since the mid-1990s, southeast Australia — where most Australians live — has experienced a 15% decline in late autumn-early winter rainfall and a 25% decline in average rainfall in April and May. In January 2020, Australia faced a severe drought, spurred by three successive winters with very little precipitation. Dry weather, high temperatures, and wildfires go hand-in-hand.

The heightened intensity and frequency of wildfires accords with scientists' predictions for a warming world. Extreme heat and volatile conditions can exacerbate early season burning, not normally seen in Australia. They can be considered a function of climate change.

Even celebrities are coming on board. Australian actor Russell Crowe skipped the January 5, 2020, Golden Globes awards ceremony because of the blazes (his home had been damaged by the fires in November). After he won a Best Actor award for his role in *The Loudest Voice*, his message was read: "Make no mistake. The tragedy unfolding in Australia is climate change-based."[18]

---

[18] https://www.bbc.com/news/world-australia-51003814 "Russell Crowe sends fires climate message to Golden Globes," *BBC News*, January 6, 2020.

There were warnings. In an advice issued in November 2019, Australia's National Environmental Science Program was unambiguous. "Human-caused climate change has resulted in more dangerous weather conditions for bushfires in recent decades for many regions of Australia."

Despite a determined political smokescreen, scientists are in no doubt that global heating has contributed to Australia's fire emergency. During that long, hot summer, Graeme Readfern in *The Guardian* noted how the Liberal Party spokesperson deflected and denied climate change as Australian bushfires raged wildly. Ancillary factors such as hazard reduction and arson were cited out of proportion to their actual effects. Prime Minister Scott Morrison continued to resist linking climate change and bushfires to any emissions reduction method, preferring inaction, defense of coal and other fossil fuels, going slow on the global UN Paris Agreement guidelines, and reluctantly cutting short his Hawaiian vacation.

Thousands of protesters took to the streets in Sydney, Melbourne, and other cities across Australia on January 10, 2020. Demonstrators called for an end to fossil fuel subsidies and action on climate change, and they shut down some roads while demanding that Prime Minister Morrison leave office.

"We're probably looking at what climate change may look like for other parts of the world in the first stages in Australia at the moment," Professor Chris Dickman summed up the continent's traumatic, tragic start to 2020.

Climate Change and Wildfires in California

Large-scale forest fires can happen anywhere in the United States. In 2019, the National Interagency Fire Center based in Boise, Idaho, coordinated emergency logistical responses to major wildfires in 30 states. During 2011–2020, an average of 7,520,000 acres have burned annually across the country, taking thousands of homes and causing dozens of deaths each year.

In this past decade, accidental fires in the American West and Southwest have become increasingly destructive of private property, especially in California. This increase is only partially due to the number and extensiveness of the fires; it is also due to a growing population of humans willing to build and buy valuable homes in highly risky areas. As summarized by Jason Metz, a former insurance investigator now on the editorial staff at *Forbes Advisor*, of the 10 American metropolitan areas most at risk of catastrophic damage from wildfires, seven are in California, two are in Colorado, and one is in Texas.[19]

According to Metz, 15% of forest fires are sparked by lightning. The other 85% are caused by people being careless with cigarettes, campfires, brushfires, and gasoline-powered tools as well as electrical equipment. In addition to these common causes, every year there are a few unusual cases of deliberate arson by criminals and neglected maintenance by power utilities.

All firefighters learn that the three factors essential for fire are air, fuel, and heat — specifically enough heat energy to ignite a flash point and sustain an oxidation chemical reaction. Efforts to explain the apparent spike in forest fires have led researchers to consider all three factors as possibilities to be investigated. In other words, although the proximate cause of most wildfires may be human carelessness, the larger context creates changing conditions regarding air, fuel, and heat that gradually make wildfires more severely threatening.

Regarding air, the concentration of oxygen in the earth's atmosphere is constant at 21%, but wind patterns blowing counterclockwise from high-pressure systems over the interior Great Basin east of the Sierra Nevada Mountains have been peaking in recent years — driving downhill Santa Ana winds in southern California and Diablo winds in northern California.[20] Regarding fuel, increasing damage by insects, particularly the mountain pine beetle, is adding to the standing deadwood and

---

[19] https://www.forbes.com/sites/marshallshepherd/2017/10/11/the-science-of-diablo-winds-fanning-california-wine-country-fires/?sh=22c137452751, "The Science of the 'Diablo Winds' fanning California wine country fires," Marshall Shepherd, *Forbes*, October 11, 2017.

[20] https://www.forbes.com/sites/marshallshepherd/2017/10/11/the-science-of-diablo-winds-fanning-california-wine-country-fires/?sh=22c137452751, "The Science of the

forest-floor litter. Entomologists agree that the recent lack of hard winter freezes has allowed the beetle population to explode, resulting in the killing of millions of trees. Regarding heat, environmental scientists conclude that global warming is the most significant contributing factor.

*****

It is important to realize that some species of plants depend on fire to fulfill their life cycle. For example, certain conifer trees do not release their pinecone seeds until scorched by fast-moving flames. In this sense, it is accurate to say that they have co-evolved with fire. In prehistoric times when Native Americans noticed that lightning started fires, they learned to start their own, carefully planning the scope and scale of the burn to alter forest habitat to their own advantage. These intentional burns continued for over 5,000 years until they were outlawed by the California government in the 19th century.

In the February 2020 issue of *Nature Sustainability*, Rebecca Miller and Chris Field of Stanford University and Katherine Mach of University of Miami published a rigorous technical study proposing a return to controlled burns as a way to reduce the deadwood problem.[21] The authors acknowledged, however, that existing California laws as well as lack of funding and trained personnel rendered such a solution unlikely. To their dismay, politically conservative columnists used the article as an opportunity to fix blame for wildfires on California Democrats, due to poor land management policies.

Miller, Mach, and Field replied to the Republican pundits and at the same time a more general readership in *Scientific American*: "The science is clear. Climate change plays an undeniable role in the unprecedented

---

'Diablo Winds' fanning California wine country fires," Marshall Shepherd, *Forbes*, October 11, 2017.

[21] https://www.nature.com/articles/s41893-019-0451-7, Miller, R. K., Field, C. B., and Mach, K. J., Barriers and enablers for prescribed burns for wildfire management in California, *Nature Sustainability*, 3, 101–109 (2020), https://doi.org/10.1038/s41893-019-0451-7.

wildfires of recent years. More than half of the acres burned each year in the western United States can be attributed to climate change. The number of dry, warm, and windy autumn days — perfect wildfire weather — in California has more than doubled since the 1980s."[22]

Although California meteorologists used to speak of a fire season in late fall, the proportion of the blustery, fire-conducive days of low humidity has risen around the calendar to establish the threat of wildfires as a constant phenomenon. These year-round fires are mostly small outbreaks, however, that are quickly extinguished.

*****

The California Department of Forestry and Fire Protection (CALFIRE) and the National Interagency Fire Center (NIFC) maintain lists related to California wildfires.[23] California is approximately 100 million acres, and since 2000, the area that burns annually has ranged between 0.1% and 1.6%. The three lists are largest wildfires, deadliest wildfires, and most destructive wildfires. The lists document that wildfires are nothing new in California. As three examples, the Santiago Canyon Fire of 1889 burned 300,000 acres, the Berkeley Fire of 1923 destroyed 640 businesses and homes, and the Griffith Park Fire in Los Angeles County in 1933 killed up to 58 firefighters.

Nevertheless, the damage done in recent years is alarming. The Atlas Fire in October 2017 burned 781 structures and killed six people. The Tubbs Fire, also in October 2017, burned 5,643 structures and killed 22. The Camp Fire in November 2018 burned 18,804 structures and killed 85, including 51 in the town of Paradise. The fire known as the LNU[24] Lightning Complex in August 2020 burned 1,491 structures and killed six,

---

[22] https://www.scientificamerican.com/article/climate-change-is-central-to-californias-wildfires/ "Climate change is central to California's wildfires," Rebecca K. Miller, Katherine J. Mach, and Christopher B. Feld, October 29, 2020.

[23] https://www.fire.ca.gov; https://www.nifc.gov/.

[24] The LNU refers to the local unit of the California Department of Forestry and Fire Protection (Cal Fire), which is named the Sonoma-Lake-Napa Unit. See https://www.ca.gov/agency/?item=california-department-of-forestry-and-fire-protection

while the North Complex Fire, also in August 2020, burned 2,320 structures and killed 15. During the 2020 wildfire season, over 8,100 fires contributed to the burning of nearly 4.5 million acres of land, making it the largest fire season in recent times.

The burning of 4.5 million acres comprises roughly 4.5% of the total land area in the state, or about three times the expected maximum of 1.6% based on the experience of the past 20 years. During the 1950s and 1960s, the burn area was about 250,000 acres, or approximately 0.0025% of California land per year. Although it may seem that this large increase over a span of 50 years represents an unsustainable calamity, it is actually a return to previous levels.

Prior to 1850, about 4.5 million acres burned yearly, with wildfire activity peaking roughly every 30 years, when up to 11.8 million acres of land burned. The much larger wildfire seasons in the past can be attributed to Native Californians regularly setting controlled burns and allowing natural fires to run their course.

The estimate of 4.5 million acres burning annually before the Gold Rush in 1849 was reported by Paul Rogers in the *San Jose Mercury News* of August 23, 2020. His article, titled "California fires: State, feds agree to thin millions of acres of forests,"[25] is a profile of Scott Stephens, Professor of Fire Science in the Department of Environmental Science, Policy and Management, at the University of California at Berkeley, and his work at the Fire Sciences Laboratory, also known as the Stephens Lab for Research and Education in Wildland Fire Science.[26]

Case studies conducted by Stephens and his students in the Sierra Nevada Mountains have documented that mechanical thinning and managed fire does not damage populations of other woodland creatures such as mammals and birds. To the contrary, controlled thinning and burning can reinvigorate entire forest habitats. His faculty research description states that he "is interested in the interactions of wildland fires and

---

[25] https://www.mercurynews.com/2020/08/23/california-fires-state-feds-agree-to-thin-millions-of-acres-of-forests/, "California fires: State, feds agree to thin millions of acres of forests" Paul Rogers, *Mercury News*, August 23, 2020.

[26] https://nature.berkeley.edu/stephenslab/, Stephens Lab: Research & Education in Wildland Fire Science.

ecosystems. This includes how prehistoric fires once interacted with eco-systems, how current wildland fires are affecting ecosystems, and how future fires, changing climates, and management may change this interaction. He is also interested in forest and fire policy and how it can be improved to meet the challenges of the next decades, both in the U.S. and internationally."

*****

Scott Stephens at UC Berkeley, like his colleagues at Stanford University, is finally succeeding at changing public opinion and political will about the beneficial potential of strategic burning on public lands. His interest in improving fire policy internationally is a reminder that the upsurge in large fires is a global phenomenon. Bruce Lieberman, writing for Yale Climate Connections, provided an overview of the problem from an international viewpoint in July 2019.[27]

Currently, more than 2.7 million people in California live within zones of very high fire hazard. Returning to the analysis by Jason Metz in *Forbes Advisor*,[28] insurance industry statistics show that in the U.S., 99% of loss claims originate in California. Some companies are declining to write new policies in the state or are insuring only highly priced properties. Many states have adopted a Fair Access to Insurance Requirements (FAIR) plan, which are pooled risk associations among underwriters. Such coverage is expensive and difficult to acquire. In California, a home-owner may apply for insurance from a FAIR plan only after being rejected by three individual companies.

Such disincentives may eventually change the behavior of people choosing to live in high-risk fire zones. In the meanwhile, advice toward reducing the risks of fire damage includes installing fire-resistant roofing,

---

[27] https://yaleclimateconnections.org/2019/07/wildfires-and-climate-change-whats-the-connection/ "Wildfires and climate change: what's the connection?" Bruce Lieberman, *Yale Climate Connections*, July 2, 2019.

[28] https://www.forbes.com/advisor/homeowners-insurance/wildfires/ "What to know about wildfire insurance" Jason Metz, *Forbes Advisor*, May 11, 2021.

trimming overhanging tree limbs, clearing vegetation around the perimeter of dwellings and outbuildings, and having an escape plan. The ultimate futility of such measures suggests that the only reliable firebreak will be the eventual reduction of greenhouse gases in the atmosphere.

Update: Worldwide Wildfires 2022

Climate change continues to exacerbate fire conditions around the world through increasing heat waves and droughts.[29] On all inhabited continents, fires are wreaking havoc on homes, businesses, and wildlife, as well as worsening air quality. The latter especially threatens people already compromised with pulmonary-cardiac health issues.

California still reigns as an epicenter of wildfire activity, with fires near Yosemite National Park threatening the giant sequoias there. Fire outbreaks are increasing in nearby states as well. As of August, 2022, nine fires were or were recently burning in the Mt. Hood National Forest region in Oregon. Other Western States experiencing fires include Arizona and New Mexico. To the North, fires have been blazing in Canada.

Beyond North America, heat and drought in Europe have brought fires to Czechoslovakia, France, Greece, Italy, Portugal, and Spain in 2022. Fires even broke out in England during the heat waves there. In South America, Argentina has experienced fires as well.

Worldwide, wildfires are projected to increase by more than 50% by the end of this century if greenhouse gas emissions continue unabated. Cuts in the GHG emissions that are causing the Earth to warm could limit the harm brought on by these destructive wildfires. Reducing their damage will be a significant benefit to reducing GHG emissions.

---

[29] https://climateyou.org/2022/08/16/our-take-global-catastrophic-wild-fires-climate-change-by-climateyou-senior-editor-abby-luby/".

Climate Change Pushing Species Extinction

Climate change is contributing to the rise in mass mortality events (MMEs). A study by the U.S. National Academy of Sciences found 727 such events involving more than 2,400 animal populations since 1940.[30] They have occurred around the world, afflicting saiga antelopes in Kazakhstan, flying foxes (known as fruit bats) in Australia, as well as sardines, anchovies, and starfish along the Atlantic coast.[31] Disease and starvation, triggered by the altered temperatures and humidity levels of climate change, cause about a quarter of the die-offs. Climate change alone another quarter of them, and human-caused conditions like pollution about a fifth of these MMEs. The rest appear to be due to a rise in diseases, biotoxicity, and events caused by multiple stressors including starvation.

The frequency and severity of mass mortality events seem to be increasing year by year. Scientists expect additional species to be impacted, including reindeer and elk. MMEs can push any species toward extinction. Since climate is implicated in about half of all MMEs, the best preventive measure humans can take is to reduce the emissions of the greenhouse gases that cause most climate change.

---

[30] http://www.pnas.org/content/112/4/1083, Fey, S. B. *et al.*, Recent shifts in the occurrence, cause, and magnitude of animal mass mortality events, *Proceedings of the National Academy of Sciences of the United States of America,* 112 (4) 1,083–1,088 (2015); https://doi.org/10.1073/pnas.1414894112

[31] https://www.theguardian.com/environment/2018/feb/25/mass-mortality-events-animal-conservation-climate-change?CMP=Share_AndroidApp_Gmail/ "The terrifying phenomenon that is pushing species towards extinction," David Derbyshire, *The Guardian*, February 25, 2018.

Famine and Food Security

Chef Yaowadee Chookong in Chiang Mai (Thailand, 2014)

Climate Change At Our Dinner Table

We often think of climate change as something distant from our everyday lives. But climate change is with us at our dinner tables every day. Why? Because our food affects climate change and because climate change affects our food.

Food and climate change may be thought of as a two-way street. Producing food affects climate change through the greenhouse gas emissions that growing, harvesting, processing, and transporting food causes. With all these processes combined, getting food to our table is responsible for about one-third of total human-caused GHG emissions.[32]

Worse, the rate at which the food system contributes to the heat-trapping gases emitted by human activities globally is increasing each year. If those emissions continue to increase at their current pace, meeting the Paris Agreement's well below 2.0° Celsius (3.6° Fahrenheit) goal would be impossible even if non-food system emissions fell to zero today.[33]

One big source of GHG emissions is the clearing of forests to create fields. As the downed trees decay and soils are denuded, they release the carbon dioxide (CO_2) that they had absorbed from the air. Other sources of emissions include the use of fertilizers, which release nitrous oxide (N_2O), and the use of fossil fuel-driven (and CO_2-emitting) tractors and other farm machines to plant, tend, and harvest the crops. Rice paddies produce methane (CH_4), as do beef and dairy cattle and their manure. Off the farm, fossil fuels are also used to dry, store, cool, transport, and package produce.

In the other direction, climate affects food through exacerbated heat, drought, heavy rain, and floods. Climate change is already creating significant risks to the food system, with rising temperatures and changing weather patterns threatening enormous damage to crops, supply chains,

---

[32] https://www.ipcc.ch/srccl/chapter/chapter-5/, Mbow, C. and Rosenzweig, C. *et al.*, Food Security, *Climate Change and Land: an IPCC special report on climate change, desertification, land degradation, sustainable land management, food security, and greenhouse gas fluxes in terrestrial ecosystems*, The Intergovernmental Panel on Climate Change.

[33] https://www.energypolicy.columbia.edu/sites/default/files/pictures/FoodandClimate-Infoguide-CGEP_v2G.pdf, Food and Climate Change Infoguide, Columbia University Center on Global Energy Policy, May 2021.

and livelihoods. These impacts have begun to occur already, and they are projected to get much worse in the decades ahead.

Farmers are already dealing with a changing climate that is altering planting and harvest dates (the growing season). The changes are bringing high temperatures during critical growth stages and causing more heavy downpours that beat down crops and waterlog soils, more droughts in some regions, and more pest infestations.

What we eat also plays a role in climate change because most animal-based products have higher total greenhouse gas emissions than do plant-based foods. Beef production generates high emissions of heat-trapping gases per pound of food because cattle produce methane, a powerful greenhouse gas, as part of their digestive process. Their manure also produces large quantities of methane and nitrous oxide.

According to recent estimates, producing a pound of beef releases about 130 pounds CO_2-eq, lamb and mutton about 50 pounds, and, surprisingly, cheese 46 pounds.[34] Food that comes from plants produces 10 to 50 times lower CO_2-eq emissions than most animal products.

There are also links between climate change, food, and health. Decreases in protein content and micronutrients have been found in crops grown under high CO_2 conditions. Unhealthy high-calorie diets (such as those rich in refined carbohydrates, added sugar, saturated fats, and red meat) are associated with diabetes, hypertension, and heart disease. Diets that are rich in plant foods can help to ameliorate those conditions, as well as lower greenhouse gas emissions.

However, climate change is not just about what is grown and raised on farms, eaten on our dinner tables, and how it got there — it is in our trash bins as well. About one-third of food that is produced is either lost during harvest or on its travels to the consumer. It may be thrown out by restaurants, cafeterias, and grocery stores, or discarded in our homes. Incredibly, about 8% of the world's total energy consumption is used to produce food that is lost or wasted.

---

[34] Scientists have devised a way to show the combined effect of different greenhouse gases, called "carbon dioxide equivalent," abbreviated as CO_2-eq. The measure totals the heat-trapping contributions of all greenhouse gases, based on their potential to cause global warming over the length of time they are active.

What can be done to create food systems that are both healthy for people and for the planet?

On the supply side, the main ways to reduce greenhouse gas emissions are limiting conversion of natural lands to agriculture, sequestering carbon in soils, using more efficient farming techniques that lower GHG emissions per unit of production of both crops and livestock, and developing efficient supply chain and distribution systems.

On the demand side, the main ways to reduce emissions are lowering food waste, adopting healthy and sustainable diets that are rich in plant-based foods, using lower-emission cooking technologies, and improving waste disposal systems. Better waste disposal can capture methane emissions from solid food waste in landfills and then burn it to generate energy.

Ways to improve the resilience of our food system to climate change include the development of heat, drought, and flood-tolerant crops, improved efficiency in irrigation systems, and the creation or expansion of insurance programs to protect farmers from the impacts of increased climate risks. Another key way for farmers to be prepared for climate change is to nurture biodiversity in and around agricultural production areas, and lowering reliance on single species. Shortening supply chains where possible and distributing storage facilities closer to markets can also improve flexibility and lower the vulnerability of our food system.

Climate change is indeed at our dinner tables and in our trash bins. There is much that we can do to create "dinner" that is good for the Earth as well as our bodies.

The Role of Soils

Some scientists and researchers asked whether the United Nations Framework Convention on Climate Change 21st Conference of the Parties (COP21) would omit soils from their agenda. The draft agreement for COP21 that began in Paris on November 30, 2015, did not once mention the term "agriculture." Although the 54-page draft agreement did mention food security, its failure to use the word "agriculture" downplayed the crucial relationship between food production and climate. Agriculture, these scientists felt, was being marginalized and taken out of the climate change equation.

Vandana Shiva, a noted Indian scholar and environmental activist, wrote that agri-corporations were attempting to hijack COP21 by focusing the attention on sectors of energy, heavy industry, and transport, and obfuscating the key role of agriculture in emitting greenhouse gases.[35]

For decades, burgeoning agribusiness practices have allowed increasing amounts of carbon, which is stored in the soil, to be released into the atmosphere, making agriculture one of the largest contributors of greenhouse gas emissions. Soil normally contains more than three times more carbon than the atmosphere. It is unable to be stored at its normal rate when agricultural practices that disturb the soil, such as tilling, monocropping, and over-grazing expose the carbon in the soil to oxygen, and let it escape into the atmosphere as carbon dioxide. The agro-industrial depletion of once-fertile soils, together with deforestation to make way for agricultural production, and the loss of rich wetlands drained and dried out for agricultural operations all reduce the capacity of soil to counter climate change by sequestering carbon.

The Agriculture Department of France has declared that soil degradation poses a threat to more than 40% of the Earth's land surface and that climate change is accelerating this rate of soil degradation, threatening food security. According to Jean-Francois Soussana, Scientific Director for Environment of the French National Institute for Agronomical Research (INRA), cultivated soils around the world have lost on average 50% to 70% of their carbon stock. Soussana is the developer of INRA's program for carbon sequestration in agriculture called "4 per 1,000: Soils for Food Security and Climate," that shows agriculture can and must be part of the solution to climate change.[36]

The goal of the "4 per 1,000 Program" is to increase the carbon stocks in the top 40 centimeters of soil by 0.4% (i.e., 4 per 1,000) every year. This increase would lead to soil carbon sequestration throughout the world of 3.4 billion tons of carbon on a yearly basis. Triple benefits would ensue — reduction in the rise of atmospheric carbon dioxide,

---

[35] https://www.navdanya.org/bija-refelections/, Jivad — The Vandana Shiva Blog.

[36] https://www6.inrae.fr/4p1000science/About-4-per-1000/What-is-the-4-per-1000-Initiative "What is the 4 per 1,000" initiative?

improvement in food security through increased crop production, and enhanced resilience of the agricultural system to climate shocks and stresses.

The international Slow Food movement[37] appealed to organizations attending COP21 in Paris to put agriculture on the agenda. Their appeal, entitled "Let's Not Eat Up Our Planet! Fight Climate Change," urged global leaders to understand that transforming the current system of food production, processing, distribution, consumption, and waste can have a strongly positive impact on climate change.

How Can Farming Keep Carbon in the Soil?

Carbon farming[38] is an agricultural system implementing practices that improve the rate at which CO_2 is removed from the atmosphere by converting it to plant material and organic matter in the soil. Such farming can successfully sequester carbon in the soil and expand the soil's water-holding capacity. These practices can build up the organic matter in the topsoil, which is essential for growing nutritious foods to fulfill the needs of humanity as well as storing carbon to mitigate climate change.

The Four Winds Farm,[39] about two hours north of New York City and certified by the Northeast Organic Farming Association of New York (NOFA),[40] is practicing carbon farming. Carbon farming is simply agriculture that reduces greenhouse gas emissions or captures and holds carbon in vegetation and soils. Four Winds has been practicing carbon farming for over 30 years with tillage reduction techniques that keep carbon in the soil, where it serves as organic matter to feed crops, hold moisture, and reduce runoff.

---

[37] https://www.slowfood.com/, Slow Food.

[38] https://modernfarmer.com/2016/03/carbon-farming/ "Carbon farming: Hope for a hot planet," Brian Barth, *Modern Farmer*, March 25, 2016.

[39] https://fourwindsfarmny.com/, Four Winds Farm.

[40] https://www.nofany.org, Northeast Organic Farming Association of New York.

These farming practices, if widely adopted, will help to keep the earth's average temperature from rising less than 2° Celsius (3.6° Fahrenheit). Reducing emissions from fossil fuels is not enough. We have to take carbon out of the air and store it. And where better to put it than in the soil where it contributes to healthy food production and sustainable water management.

Vertical Farming

An unlikely place for a vegetable farm is a 70,000-square-foot facility in Newark, New Jersey. But that is where the urban agriculture pioneer AeroFarms has one of its many indoor farming operations. AeroFarms has partnered with Dell Technologies to produce tasty and nutritious leafy greens, using data that track the entire process from seed germination to harvesting to packaging.

Established in 2004, AeroFarms is a leader in fully controlled indoor vertical farming.[41] By controlling every aspect of production — water, light, plant nutrition, a patented reusable cloth substrate for seeding, germinating, growing and harvesting; pest management; use of modules for easy scaling; and constant data tracking — AeroFarms produces kale, watercress, and other crops. These crops are marketed under the Dream Greens brand and sold at retailers, including Whole Foods and ShopRite.

According to the company, these indoor farming practices use 95% less water and achieve yields up to 390 times that of traditionally grown crops. In July 2019, AeroFarms raised $100 million to expand its operations. In 2020, it signed a deal to build a 90,000-square foot vertical farm in Abu Dhabi, the capital of the United Arab Emirates, that will be the world's largest indoor farm.

---

[41] https://www.delltechnologies.com/en-us/customer-stories/aerofarms.htm, "A harvest full of insights," AeroFarms, Dell Technologies.

Vertical farming, also called indoor farming or urban farming, has positive implications for the climate, especially since it produces higher yields per unit area, thus freeing up rural land for reforestation, thereby storing more carbon in trees. Many start-ups see the potential for almost unlimited expansion of vertical farming given an ever-growing population that will need to be fed.

Despite some slowing of global population growth rates, the UN still projects world population to grow to 9.7 billion in 2050, from the 2019 population of 7.7 billion. Feeding everyone with healthy and sustainable food requires that current agricultural practices be transformed by the massive application of science and technology. However, while the field of vertical farming has attracted considerable investor interest, scaling what are essentially pilot urban agricultural ventures to an alternative new-world scale faces several obstacles.

One is that indoor farming is energy intensive. The sun is free; artificial light is not. If the energy for the grow lights is generated from fossil fuels rather than renewables, some of urban agriculture's climate advantage is dissipated. Another hurdle to overcome is the range of crops amenable to being grown either hydroponically or aeroponically is currently limited to a few leafy greens, although experimentation is ongoing. Although rural farmland is not free, it is cheaper than even unused urban real estate. And even with much higher yield per square meter, the cost of urban farming is an issue. A lot of vertical farms would have to be built to supply a city, and to date, nobody has come close to that goal.

For now, urban farming fills a niche market. However, scaling up what are essentially small pilot urban agricultural experiments such as AeroFarms to a world scale will require social, economic, and political transitions, as well as science and technology transformations.

Were it to scale up to the point where it competes with rather than complements traditional rural farming, urban farming might contribute to the forces driving rural smallholders from the land to the rapidly growing cities. As global temperatures rise, farmers using traditional methods on traditional cropland will see their yields decrease. Crops and farming methods have co-adapted over the centuries of stable climates. Now that climate is changing, the range of a crop's growing zone changes as temperature and rainfall patterns change. If the farmer does not adapt, yield falls, or fails entirely.

Urban agriculture can supply some of the shortfall. Whether it can scale up fast enough to feed the city dwellers remains to be seen. It will do little for the displaced smallholder farmers who migrate to the cities.

The practice of vertical farming in urban areas will undoubtedly grow. However, rural farming must change and adapt to the changing climate conditions as well. As the world urbanizes, industrial, high-carbon, high-fertilizer, high-pesticide, monoculture agriculture must transform itself as well to be more climate friendly. In the future, there will need to be a transformative shift to practices that emit less greenhouse gases, that are less polluting of soil and water, and that emphasize sustainability.

Cities and Power Outages

The view from 7 World Trade Center (New York City, 2009)

Electricity Cuts and the Hurricanes of 2017

It was estimated that Puerto Rico's 3.5 million people were without power after being hit in 2017 with Hurricane Maria, a Category 5 hurricane. The recent surge of hurricanes and tropical storms are related to climate change because of the accompanying warmer ocean and air temperatures. Puerto Rico's Governor Ricardo Rosselló and San Juan's Mayor Carmen Yulín Cruz expected it to take months or even over a year to fix the outage. Hurricane Maria also left the island of Dominica without any power at all.[42]

Over a few weeks in 2017, three massive hurricanes (Harvey, Irma, and Maria) were fatal and caused widespread destruction. Power outages caused by Hurricane Irma in the U.S. almost exceeded those caused by Hurricane Sandy in 2012, which affected over 8 million customers and about 20 million people altogether.

To avoid these catastrophes, there are advantages to building underground energy transmission systems.[43] An underground high voltage direct current (HVDC) network would safeguard electricity from destructive storms, events that are becoming more frequent and that reflect the effects of climate change.[44] For many people in Puerto Rico after Hurricane Maria, living without electricity meant no food, no running water, and no fuel, which proved fatal to many.

Halfway through the normal hurricane season in 2017, the Caribbean and the U.S. Gulf Coast had experienced the devastation caused by Hurricanes Harvey, Irma, and Maria. The connections between climate change and these storms are becoming more evident.[45] Because of global warming and the rise in temperature in the air and water, climate scientists

---

[42] https://people.com/human-interest/devastating-photos-coming-out-of-puerto-rico-show-hurricane-marias-fury/ "Remembering Hurricane Maria: Devastating photos show the storm's fury in Puerto Rico," Erin Hill, *People.com*, September 20, 2017.

[43] https://www.washingtonpost.com/news/capital-weather-gang/wp/2016/06/02/save-the-climate-and-protect-america-build-an-underground-energy-interstate-now/ "Save the climate and protect America: Build an 'underground energy interstate' now," Alexander MacDonald, *Washington Post*, June 2, 2016.

[44] http://climate.org/, Climate Institute.

[45] https://insideclimatenews.org/topic/climate-change

have linked these developments to bigger storms with stronger winds and massive amounts of rain, resulting in devastating conditions.[46] Scientists are saying storms like Hurricanes Harvey, Irma, and Maria are the new norm. Hurricanes Fiona and Ian in 2022 proved them right.

The situation was still dire in Puerto Rico, where almost all the local power lines had been downed by the 155 miles per hour winds of Hurricane Maria.[47] The heavy winds also knocked out radar, weather stations, and cell towers. The hurricane somehow avoided the island's power plants, but knocked out the 2,470 miles of transmission lines, and the almost 31,500 miles of shorter lines that also transmit electricity to some 3.4 million people.[48] The Puerto Rico Electric Power Authority (PREPA) said that most local power lines have been destroyed. The PREPA had been running an antiquated grid that it had not been able to afford to bring up to date. The damage caused by Maria has been estimated at $30 billion. For Puerto Ricans, the only source of electricity came from generators for weeks and months.[49]

During Hurricane Irma on September 10, 2017, more than 4.4 million customers in Florida were without power due to downed power lines.[50] Since then, residents like those living in Coral Gables, Florida, are considering a multimillion-dollar plan to put their power lines underground for protection against future storms.[51]

---

[46] https://www.nesdis.noaa.gov/, National Environmental Satellite Data and Information Service, Department of Commerce, National Oceanic and Atmospheric Administration.

[47] https://www.bloomberg.com/news/articles/2017-09-19/hurricane-maria-heads-for-puerto-rico-after-dominica-strike "Maria latest threat to Puerto Rico after $1 billion Irmat hit," Brian Sullivan and Ezra Fieser, *Bloomberg*, September 20, 2017.

[48] http://wfla.com/2017/09/26/satellite-images-show-how-widespread-power-outages-are-in-puerto-rico/ "Satellite images show how widespread power outages are in Puerto Rico," Colleen Seitz, *WFLA*, September 26, 2017.

[49] https://www.reuters.com/article/us-puertorico-debt-prepa/puerto-rican-power-utility-files-for-bankruptcy-idUSKBN19O02F "Puerto Rican power utility files for bankruptcy," Lauren Hirsch and Nick Brown, *Reuters Business News*, July 3, 2017.

[50] https://weather.com/storms/hurricane/news/tropical-development-florida-heavy-rain-threat-late-september-2017 "High Surf, Coastal Flooding, Rip Current and Beach Erosion Threats Continue Along Southeast Coast as Invest 99L Fizzles," Brian Donegan, *The Weather Channel*, October 1, 2017.

[51] http://grist.org/article/hurricanes-keep-bringing-blackouts-clean-energy-could-keep-the-lights-on/ "Hurricanes keep bringing blackouts. Clean energy could keep the lights on," Amelia Urry, *Grist*, September 22, 2017.

Because of the likelihood that stronger hurricanes are due to climate change, many people are now convinced that protecting power lines is necessary. According to the *Washington Post*, a majority of Americans believe the severe hurricanes experienced in 2017 are because of climate change. A *Washington Post–ABC News* poll revealed a large increase in the number of people seeing this connection,[52] when compared with the number in a previous 2005 Post–ABC poll, taken just weeks after Hurricane Katrina.

The recurring story of power outages following severe storms has prompted the Climate Institute,[53] located in Washington D.C., to study high voltage direct current (HVDC) infrastructure and propose an underground grid connecting renewable energy supplies throughout the continental U.S.[54] As climate change promises more intense hurricanes and storms pummeling the coasts, it is time to get serious about burying all those power lines. Doing so will prevent them from putting people's daily lives and the nation's economy at risk of disruption again and again.

Increased Flood Risks in Coastal Cities

Scientists at the University of South Florida's College of Marine Science,[55] together with other scientists from the University of Maine and the University of Siegen in Germany, have researched the connection between rising sea levels and increasing flood risk to cities. Focusing on the Atlantic and Gulf coasts, they studied the cities in these heavily populated coastal regions where nearly 40% of the U.S. population resides (as of 2014).[56] Using data as far back as the 1950s helped determine that the risk

---

[52] https://www.washingtonpost.com/news/energy-environment/wp/2017/09/28/majority-of-americans-now-say-climate-change-makes-hurricanes-more-intense/ "Majority of Americans now say climate change makes hurricanes more intense," Emily Guskin and Brady Dennis, *Washington Post*, September 28, 2017.

[53] http://climate.org/, Climate Institute.

[54] http://cleanandsecuregrid.org/2017/01/02/east-coast-offshore-cable-feasibility-study/, East Coast Offshore Cable Feasibility Study, 2016.

[55] https://www.usf.edu/, University of South Florida.

[56] https://oceanservice.noaa.gov/facts/population.html, "What percentage of the American population lives near the coast?", National Ocean Service, National Oceanic and Atmospheric Administration, U.S. Department of Commerce.

for what scientists classify as "compound flooding" — rising sea levels, storm surge from coastal storms, high precipitation levels, and inland river flooding — was higher in these cities.[57]

Looking specifically at New York City, the researchers found that storm surges cause the worst compound flooding when high-pressure systems from Newfoundland meet moisture-laden low-pressure systems over the Mid-Atlantic, resulting in heavy precipitation. They noted, however, that rainfall contributed only minimally to the extreme flooding that accompanied Superstorm Sandy in 2012.

The report established that rising sea levels "are the main driver for increasing flood risk to American cities."[58] The researchers also substantiated that storm surges are caused by weather patterns with high precipitation, which exacerbates compound flooding potential.

The Urban Climate Change Research Network (UCCRN) assesses the impacts of these increasing risks on cities throughout the world in the Second Assessment Report on Climate Change and Cities (ARC3.2).[59] The assessment emphasizes that coastal cities are already exposed to storm surges, erosion, and saltwater intrusion. Climate change and sea level rise will exacerbate these hazards.

The ARC3.2 author team of over a hundred scientists found that expansion of coastal cities is expected to continue, with over half the global population living in cities in the coastal zone by the mid-21st century. Assessments show that the value of assets at risk in large port cities was estimated to exceed $3.0 trillion (5% of Gross World Product) in 2005. Annual coastal flood losses could reach $71 billion by 2100.

---

[57] https://www.nature.com/articles/nclimate2736 Wahl, T., Jain, S., Bender, J. *et al.* Increasing risk of compound flooding from storm surge and rainfall for major US cities. *Nature Climate Change* 5, 1,093–1,097 (2015). https://doi.org/10.1038/nclimate2736

[58] https://www.nature.com/articles/nclimate2801 Little, C., Horton, R., Kopp, R. *et al.* Joint projections of U.S. East Coast sea level and storm surge. *Nature Climate Change* 5, 1,114–1,120 (2015). https://doi.org/10.1038/nclimate2801

[59] https://uccrn.ei.columbia.edu/arc3.2, Rosenzweig, C., Solecki, W., Romero-Lankao, P., Mehrotra, S., Dhakal, S., & Ali Ibrahim, S. (Eds.). (2018). *Climate Change and Cities: Second Assessment Report of the Urban Climate Change Research Network.* Cambridge University Press.

Climate-induced changes will affect marine ecosystems near city shores, aquifers used for urban water supplies, the built environment, transportation, and economic activities, especially following extreme storm events. Critical infrastructure and precariously built housing situated in flood zones are highly vulnerable.

Increasing shoreline protection can be accomplished by building defensive structures or by adopting nature-based solutions (NbS), such as preserving and restoring coastal wetlands. Modifying structures and lifestyles to "live with water" and enhance resilience are also key adaptive measures.

For an example of how one city is responding to these increasing and compounding coastal risks, the New York City Panel on Climate Change (NPCC) works with the city government to build resilience throughout its five boroughs.[60] The City has adopted a wide range of strategies to coastal risks, in essence, taking a "portfolio approach" that includes insurance programs, social networking, engineering of hard structures, and nature-based initiatives. The Staten Island Blue Belt manages stormwater and reduces flooding while protecting natural areas at the same time. The natural areas improve the health of local waterways and raise nearby property values.

New York City Flood Prevention Fund

An additional $100 million has been added to the New York City flood prevention fund for lower Manhattan. Mayor de Blasio announced in 2015 that the funds would be added to the $15 million that was earmarked earlier as part of the first-phase flood protection design.

The city already had a $20 billion resiliency plan that is being put into action in all five boroughs and includes other plans such as the East Side Coastal Resiliency Project of Rebuild by Design.[61] The money is part of the disaster prevention program that is being overseen by the U.S. Department

---

[60] https://www1.nyc.gov/site/orr/challenges/nyc-panel-on-climate-change.page, New York City Panel on Climate Change, NYC Mayor's Office of Climate Resiliency.

[61] http://www.rebuildbydesign.org/, Rebuild By Design.

of Housing and Urban Development's National Disaster Resilience Competition, which has a $500 million fund for disaster prevention.

The widespread devastation caused by Superstorm Sandy in 2012 affected more than 60,000 New York City Housing Authority (NYCHA) residents. As of March 2015, many boiler rooms in NYCHA apartments still had not been flood-proofed, which pushed FEMA to grant $3 billion for upgrades at 33 NYCHA developments, the largest grant the federal agency has ever made.

Among the many organizations that are working with the city are the Trust for Public Land, Columbia University, and Drexel University. These organizations are planning green infrastructure improvements along the coastal areas of the city, including playgrounds, parks, and land-conservation sites to serve as buffers against flooding.

By October 2014, 4.15 million cubic yards of sand had been added to beaches across the city, and 26,000 linear feet of dunes had been protected on Staten Island. Bulkhead repairs along 10,500 linear feet had been made city-wide. In February 2015, the New York City Panel on Climate Change projected for a mid-range estimate that the local sea level in NYC will rise between 11 inches and 21 inches by the 2050s, between 18 and 39 inches by the 2080s, and between 22 and 50 inches by 2100.[62]

Recent evidence has shown that Antarctica is increasingly contributing to global sea level changes, indicating a need to better understand how this could amplify future sea level rise projections. In 2019, the NPCC presented a new upper-end, low-probability scenario that included rapid ice melt in the Antarctic (ARIM). The ARIM scenario projects 6.75 ft of sea level rise in the 2080s and 9.5 ft by 2100 for New York City. The ARIM projection takes into account recent developments in ice sheet behavior and supplements the current NPCC 2015 projections used by New York City for planning.[63]

In May 2015, Mayor de Blasio marked Earth Day by releasing the city's new sustainability plan — "OneNYC," which is short for "One New

---

[62] https://nyaspubs.onlinelibrary.wiley.com/doi/10.1111/nyas.12591, New York City Panel on Climate Change 2015 Report Executive Summary, Annals of the New York Academy of Sciences, February 16, 2015, https://doi.org/10.1111/nyas.12591

[63] https://nyaspubs.onlinelibrary.wiley.com/doi/10.1111/nyas.14008; Rosenzweig and Solecki, Eds. New York City Panel on Climate Change 2019 Report.

York: The Plan for a Strong and Just City."[64] This detailed and comprehensive plan not only focuses on preparing for the impacts of climate change but also pledges a program to tackle social and economic inequality. The climate resilience measures included in the Plan were in large part guided by the NPCC. Progress reports for OneNYC have been issued in 2019, 2020, and 2021.[65]

Resilience in Urban Communities

A picture of houses built on stilts in Dhaka, Bangladesh, on the website for the International Institute for Environment and Development (IIED)[66] is just one example of how organizations like IIED help vulnerable rural areas and urban centers adapt to climate change. Another organization that provides the knowledge base for resilience in cities is the Urban Climate Change Research Network (UCCRN).[67]

In April 2016, IIED held their 10th International Conference on Community-Based Adaptation to Climate Change (CBA10) in Dhaka. CBA conferences have been held since 2004 in Bangladesh, Tanzania, Vietnam, Nepal, and Kenya. The previous year, more than 400 participants from over 90 countries attended CBA9, and hundreds more followed conference events and interacted online.

The theme of the 2016 conference was "Enhancing Urban Community Resilience." Collaborating with IIED is the Bangladesh Centre for Advanced Studies (BCAS),[68] the International Centre for Climate Change and Development (ICCAD),[69] and the Independent University of Bangladesh (IUB).[70]

---

[64] http://www.nyc.gov/html/onenyc/downloads/pdf/publications/OneNYC.pdf, One New York: The Plan for a Strong and Just City.

[65] https://onenyc.cityofnewyork.us/

[66] http://www.iied.org/, International Institute for Environment and Development.

[67] http://uccrn.org, Urban Climate Change Research Network, Earth Institute, Columbia University.

[68] http://www.bcas.net/, Bangladesh Centre for Advanced Studies.

[69] http://www.icccad.net/, International Centre for Climate Change and Development

[70] http://www.iub.edu.bd/, Independent University of Bangladesh.

The main focus of CBA10 was to support and engage small businesses to help them become resilient to the future impacts of climate change. "Over the years the CBA conferences have tended to focus on rural communities. So we have chosen the theme for CBA10 to be on enhancing resilience of urban communities because more than half the world population is now urban," said Dr. Saleemul Huq, Senior Fellow at IIED and founder of BCAS. Cities are increasingly recognized as being on the frontlines of climate change impacts and responses, as documented in the UCCRN Assessment Reports on Climate Change and Cities (ARC3).[71]

Many of the sessions at CBA10 presented the latest resilience developments in cities and communities across the globe. Especially vital were field trips to CBA projects across Bangladesh. Participants were hosted by local communities who shared their adaptation activities. These field trips are a challenge to manage (the previous year, 200 participants visited communities around Kenya), but the feedback shows that these experiences are by far the most rewarding part of the CBA conferences.

This hands-on component is invaluable because participants can see areas heavily impacted by drought and flooding, and forested and urban areas that are striving to become more resilient. Last year, CBA9 was in Nairobi, Kenya, where field trips included visiting Lake Naivasha to see sustainable fisheries management and water harvesting, to Narok to see how communities are adapting to drought and changing rain patterns, and to Mount Kenya to see forest conservation and reforestation projects.

If New York Goes Under Water, We All Do

New York City is preparing for sea level rise, having broken ground on a $1.45 billion East Side Coastal Resiliency Project (ESCR).[72] But Manhattan is an island, and this plan only covers the Lower East Side. It

---

[71] https://uccrn.ei.columbia.edu/first-uccrn-assessment-report-climate-change-and-cities-arc3-2011-0; https://uccrn.ei.columbia.edu/arc3.2; https://uccrn.ei.columbia.edu/arc3.3

[72] https://urbanize.city/nyc/post/city-breaks-ground-145b-east-side-coastal-resiliency-project "City breaks ground on $1.45b East Side Coastal Resiliency Project," Diane Pham, *Urbanize New York*, April 16, 2021.

will not do to just save Wall Street without protecting the Upper East Side and all of the West Side including Harlem. And what about the other boroughs — Brooklyn, Queens, the Bronx, and Staten Island?

Even with coastal protection and strategic relocation, adaptation has its limits. Reducing climate change risks also depends on mitigation not just of NYC's emissions but the country's and the world's. Resilience and adaptation can only protect New York City for so long. Humankind has emitted so much heat-trapping gas into the atmosphere that even if we were to stop all greenhouse gas emissions today, sea levels would continue to rise. When the West Antarctic "doomsday" ice sheet collapses, the ocean's rise will be measured not in millimeters but in meters.[73] If we fail to act now, our children alive today will live to see New York City and most other coastal cities worldwide flooded with every full moon and passing storm. Economies will tank, real estate values plummet, millions of fleeing city-dwellers will have to relocate — but where will they go? Who will welcome them?

Ambitious as the plan is to protect Lower Manhattan from the inevitable rising seas, it can only do so much. The plan's scope needs to be expanded to cover not only the whole of Manhattan but the outer boroughs as well — Brooklyn, Queens, the Bronx, and Staten Island. Furthermore, New York is just one city of many at risk in the U.S. and the world. All cities need to develop and implement their own climate action plans.

---

[73] https://www.livescience.com/antarctic-ice-collapse.html "A third of Antarctic ice shelves could collapse at current pace of warming," Rafi Letzter, *Livescience*, April 13, 2021.

Migration, Conflict, and Population

Chuquicamata, the largest open-pit copper mine in the world (Chile, 2016)

New Wave of Climate Refugees from Puerto Rico

After Hurricane Maria, which was the strongest hurricane to hit Puerto Rico in more than 80 years, dire predictions were made to expect one of the largest mass migration flows from Puerto Rico to the United States mainland in recent history. These predictions forecast that about 200,000 storm victims would leave the island to start new lives to the north, joining other "climate refugees" from southernmost Louisiana and the shrinking islands of Alaska's Bering Strait.[74]

Climate change is on course to create more refugees and mass migrations around the world, not just in Puerto Rico. Disasters that come on suddenly, such as floods, hurricanes, and wildfires, have been reported to force an average of 21.5 million people yearly out of their homes around the world.[75] How many climate refugees will there be in the future? These claims have gained significant currency, with a common projection being that the world will have 150 to 200 million climate change refugees per year by 2050.[76]

The estimated 200,000 refugees from Puerto Rico was probably based on approximately 5% of the total population of 3.3 million in 2017. How realistic is that estimate likely to be? Very hard to say, especially given the still unknown speed, scale, and efficacy of the federal disaster relief effort. Latin family ties bind, so many Puerto Rican mainland residents will help out, either with recovery and rebuilding or support for temporary or permanent immigration.

---

[74] https://www.scientificamerican.com/article/puerto-ricans-could-be-newest-u-s-climate-refugees/ "Puerto Ricans could be newest US Climate Refugees," Daniel Cusick and Adam Aton, *Scientific American*, September 28, 2017.

[75] http://www.unhcr.org/uk/news/latest/2016/11/581f52dc4/frequently-asked-questions-climate-change-disaster-displacement.html, "Frequently asked questions on climate change and disaster displacement," UNHCR, The UN Refugee Agency.

[76] https://abcnews.go.com/Technology/wireStory/climate-change-creates-migrants-biden-considers-protections-77190887 "Climate change creates migrants. Biden considers protection," Julie Watson, *ABC News*, April 21, 2021.

Update: Several years later, some statistics were available.[77] The exact tallies may be off, as some Puerto Ricans who fled to the States eventually returned, some went and returned several times, and some put down roots in the U.S. mainland and now consider it their home. The best estimate as of late 2020 is that about 123,000 Puerto Ricans emigrated as a result of the devastation of Hurricane Maria. Many Puerto Ricans went to New York City, a magnet for Puerto Rican emigration since the 1950s, and where a large community is well established. Sizable numbers also went to other northern cities, including Chicago, Philadelphia, Newark, and Hartford. Miami was a favorite destination, given its proximity to Puerto Rico, its similar climate, and the large Cuban and other Latin American communities that make it a bilingual city.

Another factor influencing the number of climate refugees from Puerto Rico is the economic job prospects of the mainland U.S. versus the island. Long before Hurricane Maria, the economy of Puerto Rico was hurting or anemic. The government was out of funds or nearly so.

Much of the industrial base of Puerto Rico did not survive the storm. Nor did Puerto Rico's agricultural sector. Both were hit hard, and both have yet to fully recover, five years post-disaster. FEMA, the U.S. federal government's emergency management agency, was unprepared for a disaster on the island's territory. Its relief efforts were slow to arrive, stalled in the port, and haphazardly distributed. They were also entirely inadequate.

Restoration of electricity to the populace, an obvious priority, got off to a poor start when FEMA awarded a no-bid contract to a politically connected firm. Much of Puerto Rico's population was without electricity for six months to a year, and even after two years, about 20% of Puerto Ricans were still in the dark and had to rely on generators to provide light, refrigeration, and for some handicapped people, live-giving oxygen. The "relief" provided through FEMA brought few benefits for small businesses or small farmers.

The long, slow economic recovery pushed many to emigrate. The island's smallholder agriculture has rebounded only slowly; larger, more industrialized and better capitalized farming operations have recovered to a greater extent.

---

[77] https://www.census.gov/library/stories/2020/08/estimating-puerto-rico-population-after-hurricane-maria

More light-industry tax-free zones in food or clothing processing and finishing could be set up to help the island's economy to recover. Island economies face tough constraints. Many products and raw materials have to be imported by sea or air, and finished products exported, both of which raise costs.

Tourism, a principal contributor to Puerto Rico's gross domestic product (GDP), has recovered more quickly than other parts of its economy. It had been losing its allure as a vacation destination due to lackluster branding and under-investment. However, after Maria, the industry responded relatively quickly to exploit the island's natural possibilities for tourism development. The island has the appeal of tropical weather, beautiful beaches, and the charming district of Old San Juan. For East Coast Americans, it has advantages of relative proximity. Travelers from the U.S. have the convenience of no currency conversion needed, and no special visa nor passport is required. English is widely spoken. Though the tourism industry will take years to be fully revitalized, it has rebuilt significantly, which certainly helps Puerto Rico's long-term economic viability.

However, tourism's progress stalled because of the COVID-19 pandemic. Hurricane Maria dealt a body blow to Puerto Rico's health infrastructure. Many clinics and hospitals were either destroyed or rendered unable to provide care due to the lack of electricity or gasoline to run generators. Drugs and medicines ran out, and resupply was delayed or impossible. In addition, many medical personnel opted to seek work in the U.S., leaving Puerto Rico poorly prepared to deal with the pandemic.

To contain the spread of the virus, the Governor issued strict lockdown orders. For two months in early 2020, all non-essential businesses were closed, and people had to stay at home from 9 p.m. to 5 a.m. except to buy food or medicine. When these rules were relaxed, cases rose. Many were attributed to passengers arriving for vacations, so the Governor required all travelers to be tested before travel. This policy has reduced arrivals, and also new cases, which remain relatively low compared to many parts of mainland U.S. Puerto Ricans have a lot of experience with natural disasters. They have shown great resilience in weathering both climate extremes and the pandemic, but in September 2022 Hurricane Fiona knocked out the island's power yet again.

Climate Change Driving Emigration

Immigrants seeking entry to the United States have been headlining the news lately. Who are they? Where are they coming from? And why?[78]

Central America has been under increasing climatic stress for a decade. The current immigrants come mostly from Honduras, El Salvador, and Guatemala, the so-called Northern Triangle of Central American countries immediately to the south of Mexico. The stable conditions of sun and rain needed for subsistence agriculture have broken down, in part because of the natural cycles of El Niño and La Niña, and in part due to climate change that has brought longer, harsher droughts and fiercer, flood-causing rains. The food insecurity experienced by smallholder farmers forces them to migrate to the cities, where jobs are few and gang violence often endemic, especially in Honduras and El Salvador.

Families often move first to another city, then a neighboring country, then to the U.S. This multi-stage movement reflects a growing level of desperation as their food insecurity increases.

The flow of immigrants from Latin America to the U.S. is augmented by emigrants from Asia, the Middle East, Africa, and the Caribbean. Many fly to Ecuador or Brazil, which have lax visa requirements. They then travel northward, often via the Pan American Highway. However, one section of that road, between Columbia and Panama, remains unbuilt. Called the Darien Gap, it is controlled by lawless gangs who prey upon the emigrants travelling north, robbing them of whatever they have of value. It is a perilous and sometimes deadly stretch.

A largely unrecognized and unacknowledged driver for emigration is climate change. Viable solutions for the Latin Americans lie in part in adopting resilient farming practices, and in part by bolstering the remittances from relatives abroad that sustain many families. Beyond the latter family-centric measure, economic development is needed to alleviate the pressure on local agriculture to provide food security. That development

---

[78] https://climate-exchange.org/2016/04/19/where-are-all-the-climate-refugees-why-we-need-to-include-climate-change-to-the-conversation-about-the-refugee-crisis/ "Where are all the climate refugees? Why we need to include climate change to the conversation about the refugee crisis," Kilian Raiser, *Climate Exchange*, April 19, 2016.

path would require more bilateral aid from multiple donor countries and more foreign investment by American or European multinational firms in the economies of the Northern Triangle countries.

As for the emigrants from elsewhere, they are also driven by insecurity, either economic or political, both of which are made worse by climate change. If climate change is not brought to a manageable level, the flow of climate refugees could become a deluge that destabilizes and topples regime after regime.

Whether humans, organized as they are into multiple sovereign nations, can summon the collective will to halt the disruptive impacts of climate change remains an open question. The U.S. could help first by curbing its own greenhouse gas emissions that contribute to climate change — which underlies the immigrant push — and also by calling these immigrants by their true name — "climate refugees." Additionally, the U.S. should provide much of the aid and investment needed to improve the local economies of the Northern Triangle. With a more humane and self-interested foreign aid policy, the U.S. could relieve much of the food insecurity in its Central American neighbors.[79]

Be Prepared for U.S. Climate Migrations

Thank goodness scientists from the University of Southern California (USC) are studying U.S. climate migration.[80] Sea levels are rising, and this change will force hundreds of thousands — if not millions — of residents of cities on the Atlantic, Gulf, and Pacific coasts to migrate inland. The study from USC assumes that future migrants will take the same actions as those who were forced from their homes by Hurricanes Katrina and Irma. Most of those migrants flocked to the inland city nearest to the one

---

[79] https://www.thedailybeast.com/climate-change-sparked-the-border-migration-crisis "Climate Change Sparked the Border Migration Crisis," Tanya Basu, *Daily Beast*, June 21, 2018.
[80] https://journals.plos.org/plosone/article?id=10.1371/journal.pone.0227436 Robinson, C., Dilkina, B., Moreno-Cruz, J., Modeling migration patterns in the USA under sea level rise, *PLOS ONE* (2020), https://doi.org/10.1371/journal.pone.0227436

they were fleeing. The researchers therefore focused on five cities: Atlanta, Houston, Dallas, Las Vegas, and Denver.[81]

All these cities have begun to develop plans for an influx of climate migrants that is expected to swell the "normal" growth of these urban magnets. The study does note that several of these cities are prone to droughts, a risk likely to increase as the climate crisis progresses. However, they do not take into account the temperature rise that will affect almost the entire U.S. With the exception of mile-high Denver, cities throughout the country will likely become ever more inhospitable places to live. Currentresults.com projects that many urban areas will experience up to six months of consecutive days when the thermometer hits or exceeds 37° Celsius (100° Fahrenheit). On many days it will spike to 46° Celsius (115° Fahrenheit).

Las Vegas already has its Strip, a stretch of connected air-conditioned hotels and casinos that make it unnecessary for tourists to venture outside into the scorching heat, but pity the poor residents who will live and work there. Atlanta has an underground shopping and entertainment complex in the downtown Five Points. It might attract more tourists and residents when above-ground conditions deteriorate. Toronto's underground PATH complex, the world's largest subterranean city, seems to be thriving (especially in the winter). Dubai has taken to air-conditioning the outdoors, but no American city has yet to seriously consider emulating the emirate.

Climate refugees will stress not just these but many cities throughout the country. Every city should develop plans to accommodate refugees with housing, employment opportunities, transportation infrastructure, energy, even open spaces. Cities must incorporate within those plans ways to deal with the uncomfortable, even killing, heat that is sure to attend climate change. This heat will disproportionately affect the poor and all-too-often homeless, the elderly, the very young, and the medically

---

[81] https://onezero.medium.com/rising-seas-may-force-u-s-climate-refugees-to-the-same-5-cities-cc596d46e8ac "Rising Seas May Force U.S. Climate Refugees to the Same 5 Cities," Drew Costley, *OneZero*, February 7, 2020.

compromised. Southern cities, regardless of their foresight in planning for the climate changes sure to come, may become less attractive residential environments. Cities in the South have been experiencing a surge in population since COVID, due to lower housing costs and the desire of people to get out of high-density areas. However, as climate change progresses, industries will likely seek more conducive sites for their businesses. Amenities and services are likely to follow.[82]

Most American climate migrants will not relocate to a nearby inland city as did their predecessors forced out by Katrina and Irma. They will instead seek refuge not only from rising seas and increasingly frequent severe storms but also from the evermore debilitating heat. This heat will accompany and characterize the climate changes wrought by our addiction to the combustion of fossil fuels. Migrants will head for Northern tier cities — Seattle, Minneapolis-St. Paul, Boise, Des Moines, Milwaukee, Chicago, and in the East — Syracuse, Albany, and Bangor, to name just a few. Some will attempt to immigrate to Canada, visas permitting — to cities such as Toronto, Montreal, Ottawa, and Vancouver. Others will choose Alaska — Anchorage, Fairbanks, or Nome. The watchword for Northern cities that have been largely unaffected to date by the climate emergency is "be prepared," or a flood of people will inundate you.

The USC study confines its focus to American climate migrants, not encompassing the global impact of climate change or the push and pull aspects of migration. Every potential migrant is facing "push" factors that are increasingly intolerable conditions such that the person feels compelled to leave home. "Pull" factors are characteristics of the destination that the person believes offer better life chances. More study is warranted on the social and economic consequences of climate-induced migration.

Europe in the last few years has been deluged by refugees from Africa seeking respite from wars or political unrest, either caused or exacerbated by the hunger and desperation induced by prolonged droughts. Such hordes of foreigners have so frightened the receiving societies that several

---

[82] https://www.currentresults.com/Weather-Extremes/US/hottest-cities.php "Hottest cities in the United States," Liz Osborn, Current Results.

of them have spawned far-right nationalistic parties that have up-ended the long-existing political order, for example, in Poland and Hungary.

The U.S. administration from 2016 to 2020 responded harshly to the perceived rise in migrants from the Northern tier of Latin American countries. It was unaware or indifferent to the fact that these people have been pushed from their homes by climate changes that have made local farmers unable to sustain their families and by economic conditions. These circumstances foster violent gangs in overcrowded cities that prey on defenseless citizens. Those forced from their homes seek a haven and a better life north of the border. So far, those braving the dangers of immigrating to the U.S. have been a trickle, but in future decades, as the climate deteriorates, they may well become a flood that no wall can contain.

America is not ready for this coming tide. The social and political implications of climate change have barely started to enter people's awareness, much less the national dialogue. This dialogue must occur, however, because we have seen the upheaval that has roiled Europe. Brexit was in part fueled by the fear of unconstrained immigration eroding British customs and values. The U.S. is a nation of immigrants, but it is not immune to xenophobia.

America's Migration Trends

The "4 Great Migrations," a column by Charles M. Blow in the February 23, 2021, issue of the *New York Times* is a fascinating and troubling delineation of four major migratory waves destined to wash over America in the next 50 years.[83] Blow describes four environmental and demographic trends already well under way.

As temperatures and sea levels rise with climate change, millions of Americans will be forced to seek cooler, less flood-prone regions to resettle. How many millions depends on how quickly we humans curb the

---

[83] https://www.nytimes.com/2021/02/21/opinion/texas-climate-migration.html, "The 4 Great Migrations," Charles M. Blow, *The New York Times*, February 21, 2021.

emissions of heat-trapping gases produced by burning coal, oil, and natural gas that are driving the global heating and sea level rise.

A second wave is both environmental and demographic. Immigrants especially from Latin America since 1965, together with their children and grandchildren, have contributed 55% of America's growth in population from 193 million then to 324 million today. That total is expected to rise to 441 million by 2070, with 88% of the increase due to immigrants and their descendants. There will be more Asian Americans than African Americans, and there will be twice as many Latinos as Blacks. The political implications will be profound.

Cities have long been magnets for rural and small-town youth. That urbanization trend had been accelerating. Millennials in greater numbers than previous generations were seeking the excitement, entertainment, and employment opportunities that cities offer. However, the coronavirus pandemic has reversed this trend somewhat due to people fleeing crowded cities and the widespread adoption of telework.

Blacks, who fled the oppressively segregated South in the Great Migration that lasted from 1915 to 1970, are increasingly moving back to the South, abandoning the now aging, rusting, and deindustrialized Northern metropolises.

Blow does not speculate on how the confluence and interplay of the four great migrations he delineates will play out. A lot depends on how bad the climate gets, how high the seas rise, how often tropical storms devastate coastal regions, how many regimes topple under the influx of climate migrants, and how much we experience the pressure of failed crops and rising food prices.

The species that put people on the moon and landed a rover softly on Mars can achieve incredible things when it summons its collective will. We have nearly made Earth uninhabitable for ourselves and much of nature. We must now subsume our many fears, differences, and animosities in order to collectively preserve the only true home humankind will ever know.

Climate Change and World Population

Climate change will almost assuredly depress world population growth, turning it negative because of rising temperatures, lower crop yields, and reduced agricultural worker productivity. Increased mortality will be brought on by higher incidence of heat stroke and other climate-related diseases. Warmer and increasingly polluted oceans and overfishing will decimate fish populations and could lead to widespread starvation for people living in maritime-dependent societies. Finally, social and political stresses will be engendered by climate refugees forced to relocate by rising sea levels, increasing droughts, and exacerbated floods.[84] In the light of these life-threatening impacts of climate change, maternal fertility decisions will surely be affected as well.[85]

The population of the Earth is now at risk due to anthropogenic climate change, not only from catastrophes such as pandemics and wars. The world is becoming a much less hospitable place for people and indeed for almost all forms of life.

Here's a thought experiment: Could humans die off as did the dinosaurs? If this now seemingly unlikely event does come to pass, what would happen? Answer: The Earth would continue without us and would eventually recover from our depredation over the course of millennia. A sobering thought.

---

[84] https://planetsave.com/2008/07/11/taking-on-population-and-climate-change/,   "RH reality check: Taking on population and climate change," Carolyn Vogel, *PlanetSave*, July 11, 2008.

[85] https://www.wired.com/story/the-world-might-actually-run-out-of-people/ "The world might actually run out of people," Megan Molteni, *Wired*, April 2, 2019.

Dimension 4

Laws and Leaders

Columbia University (New York City, 2009)

Laws and Leaders

Dimension 4: Laws and Leaders documents the barriers and bridges encountered by governments and international organizations as they work to solve the worldwide challenge of climate change. Every country and their respective citizens are involved, despite widely differing development pathways. Section 1 on "International Action" describes the multilateral policy process as it has unfolded over the last decade. (See Table 1 in Appendix 2 for a chronology.) Most countries have pledged to contribute to reaching the temperature rise-limiting targets of the historic United Nations Convention on Climate Change (UNFCCC) Paris Agreement passed in 2015, but they have been slow in reducing their GHG emissions.

Section 2 on "Ups and Downs: The Role of the United States" documents the varying United States policies under successive presidential administrations. (See Table 2 in Appendix 2 for a chronology.) In 2017, then-President Trump announced that the U.S. would withdraw from the Paris Agreement because it undermined the U.S. economy. President Biden reversed that action the day he was inaugurated, January 20, 2021, signaling that the U.S. was intent on resuming international leadership to address climate change.

Section 3, "China, India, Europe, and the Oil Producers," focuses on the climate change roles of these countries and groups of countries. China and India are two giant developing countries each with large greenhouse gas (GHG) emissions. Both are crucial players in forging and implementing the policies that are needed to stop, reverse, or even slow the rate of climate change. Oil-producing countries, especially those in the Middle East, are also essential to involve as fossil fuel providers. Europe, including the now go-it-alone UK, has consistently supported addressing climate change both at home and on the international political stage. Section 4, "Heroes," describes the role that key people — Al Gore, Greta Thunberg, and James Hansen — have played as public, galvanizing, and scientific leaders in climate change.

Dimension 4: Laws and Leaders highlights how policies, governments, and public figures affect the outcomes of climate change, in terms of both action and inaction.

International Action

Vietnamese flag on Phú Quốc Island (Vietnam, 2014)

The Green Pontiff Urges Climate Change Initiatives

In June 2015, Pope Francis called for action on climate change in his encyclical on the environment. The 182-page *Laudato Si'*,[1] or *Be Praised*, was the first encyclical emanating from the Vatican that is dedicated to climate change. Pope Francis pressed the religious community to work for an "ecological conversion" for the faithful. He warned of the perils of industrial waste and supported the increased use of renewable fuel via subsidies and energy efficiency. He stressed that the rich nations should assume responsibility for the poorer countries and should fight inequality and global poverty.[2] The Pope also advocated for a new partnership between science and religion to successfully win the battle against what he defines as human-driven climate change. He offered the following prayer:

A Prayer for Our Earth

All-powerful God, you are present in the
whole universe
and in the smallest of your creatures.
You embrace with your tenderness all that
exists.

Pour out upon us the power of your love,
that we may protect life and beauty.
Fill us with peace, that we may live
as brothers and sisters, harming no one.

O God of the poor,
help us to rescue the abandoned and

---

[1] https://www.vatican.va/content/francesco/en/encyclicals/documents/papa-francesco_20150524_enciclica-laudato-si.html Encyclical Letter, Laudato Si', of the Holy Father Francis, on Care for our Common Home, The Vatican.

[2] https://www.washingtonpost.com/news/acts-of-faith/wp/2015/06/18/what-you-need-to-know-about-pope-franciss-environmental-encyclical/ "What you need to know about Pope Francis's environmental encyclical," Christiana Z. Peppard, *The Washington Post*, June 18, 2015.

forgotten of this earth,
so precious in your eyes.

Bring healing to our lives,
that we may protect the world and not prey on it,
that we may sow beauty, not pollution and
destruction.

Touch the hearts
of those who look only for gain
at the expense of the poor and the earth.
Teach us to discover the worth of each thing,
to be filled with awe and contemplation,
to recognize that we are profoundly united
with every creature

as we journey towards your infinite light
We thank you for being with us each day.
Encourage us, we pray, in our struggle
for justice, love and peace.

Pope Francis,
Laudato Si'

Pope Francis's prayer and encyclical are indeed wide-ranging. James Martin, S.J., a Jesuit priest and editor-at-large of the journal *America*,[3] summarizes 10 key points of the broad document:

1. The spiritual perspective is now part of the discussion on the environment.
2. The poor are disproportionately affected by climate change.
3. Less is more.
4. Catholic social teaching now includes teaching on the environment.
5. Discussions about ecology can be grounded in the Bible and church tradition.
6. Everything is connected — including the economy.

---

[3] https://www.americamagazine.org/faith/2015/06/18/top-ten-takeaways-laudato-si "Top Ten Takeaways from 'Laudato Si'" James Martin, S.J., *America the Jesuit Review*, June 18, 2015.

7. Scientific research on the environment is to be praised and used.
8. Widespread indifference and selfishness worsen environmental problems.
9. Global dialogue and solidarity are needed.
10. A change of heart is required.

The encyclical was timely, leading up to the 2015 United Nations Framework Convention on Climate Change 21st Conference of the Parties (COP21) in Paris,[4] the much-anticipated conference that solidified a global agreement to limit greenhouse gas emissions.

Pope's July Meeting Instills Courage

The Pope hosted another climate meeting early in July 2018, this one not with fossil fuel and finance heavyweights but with prestigious environmental activists.[5] Such attendees can influence debate, but not policy. It is good to have the Pope make the climate a top concern, but can even he impel a recalcitrant, indifferent world into action?

Immediate measures to curb carbon emissions are required to avoid the most catastrophic damages of climate change worldwide. Not more talk. Governments must commit to policies ending subsidies for fossil fuel, closing legacy coal mines and power plants, retraining workers, and supporting affected communities. Governments must also incentivize the switch to electrical vehicles (EVs). And they must invest much more heavily in public health and social safety-net measures, finding ways to lead to a sustainable future for the human race on the planet Earth.

The fossil fuel industry must redouble efforts to accelerate the transition to renewable low-carbon energy sources. If they do not, governments must force them to do so. Financial institutions and global development agencies must stop funding fossil fuel projects. Scientists, engineers, and entrepreneurs must speed up developments in battery technology, carbon

---

[4] https://unfccc.int/process-and-meetings/the-paris-agreement/the-paris-agreement, The Paris Agreement.
[5] https://www.axios.com/pope-to-huddle-with-environmental-leaders-activists-1530207431-7cab74cb-bcde-46c5-8bdd-177588682b83.html, "Pope to huddle with environmental leaders, activists," Eric J. Lyman, *AXIOS*, June 28, 2018.

capture and storage technology, zero-emission hydrogen fuel technology, biodegradable plastic technology, urban agriculture, meatless meat, and other ways to create a sustainable planet.

Citizens of every country must become aware that the climate is changing, that they are responsible directly and indirectly for its deterioration, and that they can, by collectively demanding action to halt greenhouse gas emissions, force their governments and corporations to ensure a future for their children and for generations to come.

What must be done is clear, but it will entail change, which is always difficult. Talk must give way to concerted action. Somehow enough popular will must be aroused to force governments to find the courage and will to effect rapid, though unpopular, change, corporations to eschew short-term profits for long-term viability, and financial institutions to divest resources from extractive activities. Environmental damage must be halted and resources deployed that are required for a new, sustainable world. The Pope, through his convening power as a global spiritual leader, is doing his part to instill the courage and will required to save the planet.

World Wins, Hydrofluorocarbons Lose

Besides working together to reach the Paris Agreement, which deals solely with climate change, countries have also been collaborating on the Montreal Protocol that limits the production of hydrochlorofluorocarbons (HCFCs) and chlorofluorocarbons (CFCs). These two sets of potent heat-trapping greenhouse gases are also culprits in the opening of holes in the ozone layer. The Montreal Protocol, signed in 1987, helped save the ozone layer by nearly eradicating the use of the ozone-depleting chemicals HCFCs and CFCs.[6,7]

Ozone is a molecule that contains three oxygen atoms that are constantly being formed and destroyed. Both HCFCs and CFCs accelerate the destruction of ozone atoms, creating gaping, growing holes in the ozone

---

[6] https://ozone.unep.org/treaties/montreal-protocol, "The Montreal Protocol on Substances that Deplete the Ozone Layer," UN Environment Programme: Ozone Secretariat.

[7] Successful as the Montreal Protocol has been in bringing about protective bans, laws, and penalties, there still exists a "banked" supply in old appliances, as well as a black market in CFCs, with illegal imports smuggled in mislabeled packaging. https://www.achrnews.com/articles/97845-buying-selling-black-market-cfcs-it-s-really-not-worth-the-risk.

layer, which is concentrated in the stratosphere about 14 to 28 km (9 to 18 miles) above the Earth. The ozone layer shields the planet from ultraviolet rays, which are harmful to humans. Exposure to ultraviolet radiation can cause skin cancer, cataracts, and impaired immune systems.

Hydrofluorocarbons (HFCs), which contain no ozone-destroying chlorine, were developed to replace CFCs and HCFCs in air-conditioning and refrigeration. They are, however, also potent greenhouse gases, with high global warming potential. A phase-down of these dangerous HFCs has been estimated to prevent 8.8 gigatons of equivalent carbon dioxide $(GtCO_2\text{-eq})$[8] per year in greenhouse gas emissions by 2050.[9]

In October 2016, almost 200 countries voted to phase down the use of ozone-destroying hydrofluorocarbons, which are now commonly used in refrigerators and air conditioners. The agreement was reached in the Rwandan city of Kigali as part of the Montreal Protocol.[10] The agreement amends the Montreal Protocol of 1987 and aims to reduce global warming by 0.6° Celsius (1° Fahrenheit) by 2100 through reduction in hydrocarbons. If achieved, this compact will avoid emitting 80 billion metric tons of carbon dioxide equivalent by the year 2050. Countries started to phase down HFCs in 2019 and continue to reduce usage, leading to a projected 80% reduction in HFCs globally by 2047.

While safer chemicals are now being developed to replace HFCs, a vicious cycle has become apparent. Because the HFCs released into the atmosphere contribute to global warming, the demand for air conditioners is rapidly rising, causing more and more HFCs to escape into the atmosphere.

President Obama praised the Kigali amendment to the Montreal Protocol.[11] However, as of May 2022, the U.S. Congress has still not ratified

---

[8] Carbon dioxide equivalent or CO_2eq is the number of metric tons of CO_2 emissions with the same global warming potential as one metric ton of another greenhouse gas; it is used as a calculator for global warming potential across multiple gases.

[9] https://www.pnas.org/content/106/27/10949, Velders, G. J. M., Fahey, D. W., Daniel, J. S. *et al.*, The large contribution of projected HFC emissions to future climate forcing, *PNAS* 106 (27) 1,0949–1,0954 (2009); https://doi.org/10.1073/pnas.0902817106.

[10] http://www.climatecentral.org/news/senate-could-block-landmark-hfc-climate-treaty-20795 "U.S. Senate Could Block Landmark HFC Climate Treaty," John Upton, *Climate Central*, October 17, 2016.

[11] https://www.whitehouse.gov/the-press-office/2016/10/15/statement-president-montreal-protocol, Statement by the President on the Montreal Protocol, The White House, October 15, 2016.

the new climate change-reducing amendment, but did pledge in April 2021 to do so.

Nuclear Autumn Means Climate Change Disaster

In July 2017, North Korea tested a ballistic missile capable of reaching the United States. The U.S. reacted by conducting a missile defense test in the Pacific Ocean, adding to the escalating threats and provocations between the two countries.[12] In February 2022, Russia invaded Ukraine and implicitly threatened to use nuclear weapons.[13] All sides need to stand down. Nuclear aftermath is not an option, especially in light of a report on potential consequences of a nuclear strike and the effects on global climate patterns. A team from the University of Nebraska-Lincoln published its report in the journal *Environment: Science and Policy for Sustainable Development*,[14] expanding on previous nuclear war simulations where more than five million metric tons of black carbon particles would disperse into the stratosphere.

According to the online publication *New Atlas*, "the simulations predict not so much the catastrophic 'nuclear winter' that the 20th century Cold War panic taught us, but rather a more moderate 'nuclear autumn' scenario."[15] The sun would not be occluded to the point of plunging the entire climate into a frigid, crop-killing, season-obliterating "winter." Instead the black carbon particles would dim the sun and lower the temperature enough to reduce crop yields globally to the point that food scarcity would become prevalent, commodity prices would rise, and global grain trade would come to a halt. Trade effects occur because governments are likely to ban the export of grains needed for domestic consump-

---

[12] https://www.cnn.com/2017/07/29/politics/trump-china-north-korea-tweet/index.html, "Trumps rips on China after North Korea missile test," Brad Lennon, *CNN*, July 31, 2017.

[13] https://www.nytimes.com/2022/04/01/opinion/biden-putin-ukraine-nuclear-weapons.html

[14] http://www.tandfonline.com/doi/full/10.1080/00139157.2017.1325300 Liska, A. J., White, T. R., Holley, E. R. and Oglesby, R. J., Nuclear Weapons in a Changing Climate: Probability, Increasing Risks, and Perception, *Environment: Science and Policy for Sustainable Development* 59 (4), 22–33 (2017); https://doi.org/10.1080/00139157.2017.1325300

[15] https://newatlas.com/nuclear-autumn-climate-effects/50510/, "Just one nuclear detonation could plunge Earth into 'nuclear autumn,'" Rich Hardy, *New Atlas*, July 17, 2017.

tion. Hungry people tend to blame their governments for food system failures, so political unrest could spread and escalate unremittingly. Many regimes would teeter, some would fall.[16]

A peer-reviewed paper in the *Proceedings of the National Academy of Sciences of the United States of America* is titled "A regional nuclear conflict would compromise global food security." Published in 2020 by Jonas Jägermeyr and colleagues, the paper shows that even a limited nuclear conflict could cause persistent multiyear losses that would constrain domestic food availability and propagate to food-insecure countries in the Global South.[17] By year 5, maize and wheat availability would decrease by 13% globally and by more than 20% in 71 countries, with a cumulative population of 1.3 billion people. The authors state, "In view of increasing instability in South Asia, this study shows that a regional conflict using <1% of the worldwide nuclear arsenal could have adverse consequences for global food security unmatched in modern history."

Nuclear autumn is almost as bad as nuclear winter and a lot easier to spark. One big nuclear bomb, or a couple of smaller ones, could do it. Sobering. Frightening. According to *New Atlas*, China's most powerful nuclear weapon, a 5-megaton beast, could single-handedly send the world into "nuclear autumn" with just one detonation. China has 20 of these weapons.

The study forecasts that blasting five million metric tons of black carbon into the stratosphere would dramatically decrease global rainfall anywhere from 20% to 80% depending on the specific area. There would be a life-threatening domino effect on the total volume of agricultural production because the growing seasons would be reduced by between 10 and 40 days per year for up to five years, resulting in famine. This and other cataclysmic climate effects could kill up to a billion people, mainly in developing "food-insecure" countries.

Diplomacy is the only viable strategy for the U.S. in dealing with North Korea, Russia, and other nuclear powers, because there are no good military options. In 2017, the least bad option was to start talking again

---

[16] https://news.unl.edu/newsrooms/today/article/why-a-single-nukes-impact-shouldnt-only-be-measured-in-megatons/ "Why a single nuke's impact shouldn't only be measured in megatons," Leslie Reed, Nebraska Today, July 6, 2017.

[17] https://www.pnas.org/content/117/13/7071, Jägermeyr, J. *et al.*, A regional nuclear conflict would compromise global food security, *PNAS* 117 (13) 7,071–7,081 (2020); https://doi.org/10.1073/pnas.1919049117.

with North Korea. Chastising China was not likely to help.[18] Neither will bombers buzzing Pyongyang,[19] nor holding more joint military exercises with South Korea.

President Trump pursued the diplomatic route, confident in his ability to negotiate a deal with Democratic People's Republic of Korea (DPRK) President Kim Jong-un. Trump travelled to Pyongyang to meet with Kim personally and met him a second time in Singapore. The two leaders established a cordial, even affectionate personal relationship, but the tangible results were few. Trump cancelled some military exercises with South Korea, while Kim made some vague, unenforceable promises to denuclearize, which he has not followed through on.

The threat, the stalemate, and the volatility of a nuclear-capable North Korea confront President Biden as well. He has little choice but to pursue diplomacy too, though he is working through official State Department channels. Biden can offer relaxation of the economy-crippling sanctions imposed on the DPRK by the UN and the U.S. He can also proffer substantial development assistance.

So, the tense stand-off is likely to continue for another administration, punctuated by provocations that raise tensions and force the world to pay attention to Kim and North Korea. The tensions will probably not spark a climate holocaust. Kim is fractious but not suicidal.

Our Common Future under Climate Change

At the UNESCO headquarters in Paris during the first week in July 2015, thousands of climate and social scientists and policy experts came together for a large gathering of high-ranking climate change scientists. Their goal was to clear a path for a successful COP21 Climate Summit scheduled for Paris in December 2015. The seminal conference was called "Our Common Future under Climate Change."

---

[18] https://www.cbsnews.com/news/trump-tweet-disappointed-in-china-over-north-korea/ "Trump says he's "disappointed" in China over North Korea," Emily Tillett, *CBS News*, July 30, 2017.

[19] https://www.reuters.com/article/us-northkorea-missiles-idUSKBN1AF02K "U.S. flies bombers over Korean peninsula after North Korea missile test," James Pearson and Michelle Nichols, *Reuters*, July 30, 2017.

According to spokespersons from Our Common Future,[20] about 2,000 scientists from almost 100 countries met in some 165 sessions. Experts in the physical and social sciences led the sessions, presenting ways to reduce emissions and build resilient societies in the upcoming climate change era.

Standing in for French President Francois Hollande for the opening plenary was French Research and Education Minister Najat Valland-Belkacem. "If we don't act [on climate] in the future, we can be pretty sure there will be conflicts," said Belkacem. "Thanks to you, no one can seriously challenge the role of humans in climate change."

At the conference, scientists brought forward new projections for a wide range of climate change impacts, including both terrestrial and marine systems. Collectively, those in attendance and their teams laid the groundwork for scientific, fact-based decision-making at the upcoming UNFCCC Conference of the Parties.

Chris Field, the chairman of the CFCC15 Scientific Committee and Director of the Carnegie Institution for Science's Department of Global Ecology located at Stanford University, declared, "We are moving to a post-carbon era, where climate-change mitigation and adaptation are combined with other goals to build a sustainable future."

Scientific Committee member and Professor at the University of Oslo, Dr. Karen O'Brien, added, "This conference has shown that social and technological innovation, from individuals, communities, firms, and nations, can lead to mitigation and adaptation options that are scalable, fair, and cost-effective."

Closing remarks were made by the French Minister Ségolène Royal, who was responsible for ecology, sustainable development, and energy. "I would like you to have an impact on the COP21 negotiations," Royal told delegates. "You are free to express your ideas. You don't have to hide behind euphemisms."

After the conference, a final statement released by Our Common Future explained that "a two-in-three probability of holding warming to 2° Celsius or less will require a budget that limits future carbon dioxide emissions to about 900 billion tons, roughly 20 times annual emissions in 2014."

---

[20] https://www.commonfuture-paris2015.org/, Our Common Future Under Climate Change, International Scientific Conference.

They Did It! Historic Paris Agreement Reached

After 20 years and 2 weeks of contentious debate, delegates from almost 200 countries agreed to make climate change a priority in their nations. On December 12, 2015, several thunderous minutes of applause and cheers followed the historic vote on the agreement. From this point on, climate change will be given priority by countries all over the world. Everyone who agreed to the treaty has pledged to put into motion actions that will ultimately reduce greenhouse gases.

Representatives from 196 countries with authority delegated by their respective governments adopted an international treaty called the Paris Agreement[21] on December 12, 2015, at the United Nations Framework Convention on Climate Change 21st Conference of the Parties (COP21) in Paris. The stated goal of the treaty is to hold global temperature rise to well below 2° Celsius (3.6° Fahrenheit), preferably 1.5° Celsius (2.7° Fahrenheit), compared to pre-industrial levels. It is a legally binding international agreement on climate change. The global target itself is not legally binding, but there is an obligation for nations to regularly set national targets for their contributions to the global goal.

The Paris Agreement came into force on November 4, 2016, after 55 countries representing 55% of global emissions officially joined the treaty; 191 countries have now joined. The Paris Agreement is a landmark achievement because, for the first time, a binding agreement brought all nations into a common cause to slow climate change and adapt to its effects.

To achieve this long-term temperature goal, countries aim to curtail greenhouse gas emissions enough to achieve a climate-neutral world by mid-century. However, the countries' commitments, known as Nationally Determined Contributions (NDCs), so far are not enough to achieve the goal of keeping the global temperature rise "well below 2° Celsius (3.6° Fahrenheit)." Given current commitments, scientists predict that the Earth's temperature will increase by 3° Celsius to 4° Celsius (5.4° to 7.2° Fahrenheit) by the end of century, compared to pre-industrial levels.

---

[21] https://unfccc.int/process-and-meetings/the-paris-agreement/the-paris-agreement, The Paris Agreement.

To address this challenge, the Paris Agreement sets up a "Global Stocktake" process. The national commitments are reviewed every five years to see how the world is progressing on climate change mitigation, resilience, and finance, and to provide the opportunity for all countries to increase their ambition to limit global warming.

Slowing greenhouse gas emissions globally will cost an estimated $3 trillion to $7 trillion yearly. This figure includes funds that will pay for adapting to climate change impacts, such as building and reenforcing sea walls and mitigating its causes by developing renewable energy sources such as wind and solar. The 32-page Agreement stipulates that developed countries send $100 billion yearly to developing counterparts.[22] Since that time, pledges have been significant but fall far below that level. At the first pledging conference in 2014, developed countries committed $9.3 billion to the Green Climate Fund (GCF). At the second conference, in October 2019, commitments totaled $9.8 billion, despite no payment from the U.S. To that point, the GCF had allocated $5.2 billion to 111 projects in 99 countries.

Accountability was an issue that was fiercely debated in Paris and at all COP meetings. The pact has reporting requirements but no real means to hold nations to their pledges and commitments. Without an enforcement mechanism, the Paris Agreement depends on nations to care about their standing among the collective of their peers. Some care more than others. Most have made good faith efforts to curtail emissions, but the pace of change has been far too slow to reach the Treaty's target by mid-century.

While countries came together to accept the new climate deal, and environmentalists welcomed the long-term emissions reduction goal, progress toward reaching that goal continues to be disappointing. Developing countries want to see monetary compensation for loss and damage incurred by climate change, since developed countries are primarily responsible for the greenhouse gas emissions spewed into the atmosphere.

Hope for COP26: Increasing Ambition

The 26th Conference of the Parties (COP) to the United Nations Framework Convention on Climate Change (UNFCCC) was postponed for a year

---

[22] https://unfccc.int/resource/docs/2015/cop21/eng/l09r01.pdf, Adoption of the Paris Agreement, December 12, 2015.

because of the COVID-19 pandemic but was finally held in November 2021 in Glasgow, Scotland. This COP was a critical step towards the Global Stocktake, a process agreed as part of the Paris Agreement to assess progress and to encourage augmentation of national commitments. The goal of the Paris Agreement is to "limit global warming to well below 2°, preferably to 1.5° Celsius" (3.6° to 2.7° Fahrenheit), compared to pre-industrial levels.

It was the first time that the UK hosted the Conference of the Parties of the UN Framework Convention on Climate Change (UNFCCC).[23] The UK wanted a face-to-face meeting, which they believed would induce the assembled delegates from all the countries of the world to show "increased ambition" in setting their climate change targets and timetables. Much greater efforts are needed in order for the world to fulfill the goal of the Paris Agreement.

These conferences of the parties are like medieval fairs, with tens of thousands of people coming together to encourage action on climate change. The official part of the COPs is attended by delegations from every country, all of whom are parties to the UNFCCC Agreements. Also attending are designated observers from climate change interest groups and many media outlets. Unofficially, many non-governmental organizations (NGOs) concerned about climate change come as well, many of whom put on lively side events, host colorful booths, or gather for protests.

COP26 was important for several reasons.[24] The first is that it occurred as the world was emerging from the COVID-19 pandemic, with the urgent need and opportunity to firmly embed climate action into national plans for what is labelled a "green recovery." However, many countries are returning to fossil fuel-based energy systems or even redoubling investments in them, a course anathema to meeting the Paris Agreement goals. Therefore, COP26 had an important role to play in encouraging all nations to seize the opportunity to build back greener, with climate change front

---

[23] https://ukcop26.org/wp-content/uploads/2021/05/COP26-EXPLAINED-FINAL.pdf "COP26 Explained".

[24] https://www.euronews.com/green/2021/02/27/what-is-cop26-and-why-is-it-so-important, "What is COP26 and why is it so important?", Marthe de Ferrer, euronews.green, September 30, 2021.

and center in recovery policies and programs. Simply put, climate change must be integral to the world's economic recovery.

How can these green recovery packages help to both restart moribund economies and jumpstart climate action at the same time? The main way is for governments and international donor organizations to provide incentives to significantly scale up solar and wind farms, electric vehicles, public transport, and home electrification. Besides supporting these low-carbon, "ready-to-go" technologies, governments with the capacity to do so need to ramp up research and development on zero-emission air and sea transport, carbon capture and storage, and hydrogen power. Innovations are needed to achieve net-zero emissions by 2050, the current overall goal set by the Intergovernmental Panel on Climate Change (IPCC).

Other areas ripe for action are movement toward low-carbon food systems with more plant-based diets and "nature-based solutions" (NbS). NbS involves activities such as planting trees, restoring coastal wetlands, and increasing soil fertility by adding carbon in the form of crop residues to agricultural fields. All of these will take investments in known technologies as well as in innovative research.

Another requirement to maximize the climate impact of a green recovery is to overhaul the finance mechanisms needed to fund it, so that it will be cost-effective for public sector donors and profitable for the private sector. These were all topics that the delegates to COP26 addressed.

The next reason that COP26 was important is the growing recognition that countries simply are not doing enough to reach the Paris Agreement target of holding the global surface temperature rise to below 2° Celsius. They are not even nearly reaching the targets knows as nationally determined commitments (NDCs) to which they committed themselves to achieving at the Paris COP in 2015.[25] NDCs are the way that countries signal to their global partners in the Paris Agreement their plans to reduce emissions and adapt to climate change impacts. They are deliberately designed to not be one-size-fits-all, in order to reflect the different contexts and capacities of each country to take climate change action.

---

[25] https://www.euronews.com/2020/12/11/paris-climate-agreement-five-years-on-is-europe-on-track-to-hit-the-deal-s-targets, "Climate change: Five years on from Paris Agreement, is Europe on track to hit the deal's targets?", Luke Hurst, euronews, December 11, 2020.

The UNFCCC recently tabulated the nationally determined commitments (NDCs) some countries had submitted by the end of 2020. Extrapolating from the countries that were included in the report, the UNFCCC found that the world's combined countries were not on track to meet the Paris Agreement goals.[26] According to the Intergovernmental Panel on Climate Change (IPCC), countries need to reduce greenhouse gas emissions by 45% in 2030 compared to 2010 in order to limit warming to that level. The UNFCCC report demonstrates the urgent need to increase ambition significantly in order to fulfill the Paris Agreement targets.

A final reason that COP26 was important is that it was the first since the U.S. has rejoined the Paris Agreement after pulling out during the Trump administration.[27] John Kerry, the U.S. Special Presidential Envoy for Climate, played a large role in the discussions, as he did at the Paris COP in 2015. President Biden also attended, highlighting the return of the U.S. to leadership in the global effort to limit climate change.

For all these reasons — the need to centralize climate action in the recovery from the coronavirus pandemic, the cumulatively inadequate national commitments to greenhouse gas emission reductions and the failure to reach even those low bars, and the return of the U.S. to climate leadership — COP26 in Glasgow provided a key opportunity for ensuring that the Earth remains a hospitable home for humans everywhere.

It is imperative that all with a stake in a livable world — individuals, non-profit organizations, corporations, financial institutions, as well as governments — reflect on the consequences of failing to achieve a viable climate. It is essential to reevaluate the costs of inaction and the benefits of acting decisively now, and then recommit to a global effort that entails undergoing short-term disruption for an enduring sustainable life on Earth. All must join in "increasing ambition" to attain that quintessential goal.

---

[26] https://unfccc.int/news/climate-commitments-not-on-track-to-meet-paris-agreement-goals-as-ndc-synthesis-report-is-published "'Climate Commitments Not on Track to Meet Paris Agreement Goals' as NDC Synthesis Report is Published," United Nations Climate Change, February 26, 2021.

[27] https://www.euronews.com/green/2020/11/04/us-pulls-out-of-paris-climate-agreement-what-happens-next, "US pulls out of Paris Climate Agreement: What happens next?", Rosie Frost, euronews, November 4, 2020.

Ups and Downs: The Role of the United States

Cape Cod (Massachusetts, 2013)

Obama's Clean Power Plan

"There is such a thing as being too late when it comes to climate change," said President Obama when he announced his action plan for climate change on August 3, 2015.[28] The Clean Power Plan (CPP) detailed the goals expected to be realized by 2020 to 2030. The President said the plan was "the biggest, most important step we've ever taken to combat climate change" and one that "sets the first-ever carbon pollution standards for power plants."

The plan was buttressed by the Clean Air Act, which mandates that the U.S. Environmental Protection Agency (EPA)[29] "must regulate any pollutant that is deemed a danger to human health and wellbeing." The EPA has found that copious amounts of carbon dioxide (CO_2) emitted from automobile exhausts, power plants, factories, and industrialized farms qualified as a dangerous pollutant, a finding challenged but upheld by the Supreme Court.

The Clean Power Plan intended to decrease CO_2 emissions by 32% from 2005 levels by 2030, reducing emissions by a total of 870 million metric tons. By 2030, the health benefits projected by the plan were 3,600 premature deaths prevented and 1,700 fewer non-fatal heart attacks. Asthma attacks in children would be reduced by 90,000, and health in general would be improved, lowering the number of missed workdays and schooldays by 300,000.

Then-EPA Administrator Gina McCarthy wrote on her blog that "the Clean Power Plan is projected to cut the average American's monthly electricity bill by 7 percent in 2030." She claimed electricity would not necessarily be cheaper, but she foresaw Americans using less energy because of the energy efficiency mandates that were part of the CPP.

Before 2030, EPA projections showed electricity prices would rise modestly by 2.4% to 2.7% in 2020, but then decline by 2.7% to 3.8% in 2025, and by 7% to 7.7% in 2030, when the average American family was expected to save about $7 per month and more than $80 per year on their electricity bill. In 2030, the reduced energy usage was expected to power

---

[28] https://www.whitehouse.gov/the-press-office/2015/08/03/fact-sheet-president-obama-announce-historic-carbon-pollution-standards "Fact Sheet: President Obama to Announce Historic Carbon Pollution Standards for Power Plants," The White House, August 3, 2015.
[29] https://www.epa.gov/climate-change "Climate Change," United States Environmental Protection Agency.

30 million homes, and save consumers a total of \$155 billion from 2020–2030, according to the White House CPP fact sheet.

The plan provoked Republican opposition in Congress. Senator Mitch McConnell (R-KY) has exhorted all states with coal-fired power plants and coal mines to fight the CPP's goals and requirements. The CPP has also faced court battles where these same states argued that the plan unconstitutionally takes away states' rights. Soon after President Trump took office, he effectively annulled the plan. However, Obama's Clean Power Plan did have a positive impact on the UN Framework Convention on Climate Change 21st Conference of the Parties that met in Paris in December 2015. There, 195 countries negotiated a remarkable and unprecedented compact that committed them to collectively limit global climate change.

Update: The U.S. Supreme Court curbed EPA's ability to require carbon emissions reductions in June 2022.[30]

With or Without Trump

Even though COP21 was the planet's major force to slow climate change, many were uncertain about the direction the U.S. and the Trump administration would take on climate change. Trump had campaigned to withdraw from the COP21. Trump's Environmental Protection Agency nominee, Myron Ebell, wanted to ultimately dissolve the EPA. In early 2017, Ebell spoke at an event in Brussels and called climate experts "urban eco imperialists."

Other countries, believing the U.S. would step back from its agreement in Paris, sided with China to take on climate change leadership. China, as well as India, have large investments in solar and wind power. Countries such as Germany, Britain, and France are attempting to be a part of those business investments. In early 2017, for example, the EU and China agreed to develop a platform for cooperation on energy.

At the annual World Economic Forum (WEF)[31] in Davos, Switzerland in January 2017, climate change loomed large on the agenda. Government

---

[30] https://climateyou.org/2022/06/27/scotus-poised-to-handcuff-the-epa/

[31] https://www.weforum.org/, The World Economic Forum.

officials joined business leaders and investors in realizing that sustainable business ventures were a $12 trillion opportunity and that investing in fossil fuels was no longer the way to go.

Announced at Davos was the new Green Digital Finance Alliance,[32] a group bringing together financial institutions using digital technology to advance green finance in lending, investment, and insurance. According to the International Institute for Sustainable Development (IISD),[33] Norway sponsored a fund to raise $400 million by 2030 supporting deforestation-free tropical agriculture investments. Norway committed to contribute up to $100 million.

As a major part of the Paris Agreement on Climate Change produced at the UN Framework Convention on Climate Change 21st Conference of the Parties (COP21)[34] in December 2015, almost 200 nations agreed to phase out fossil fuel use during the second half of this century. Reducing dependency on fossil fuels is essential to lessen the occurrence of floods, heat waves, and droughts and to slow rising sea levels. One of the main COP21 agreements was to create a clear line to secure funding earmarked for climate change efforts. Funding sources targeted to lower greenhouse gas emissions and for climate-resilient development are tracked by IISD.

The European Investment Bank President Werner Hoyer announced that the EIB would provide $100 billion in loans over the next five years to fund programs supporting green and sustainable development.[35] Also, Sweden announced they would phase out greenhouse gas emissions by 2045 and become a totally fossil fuel-free state, one of the most ambitious goals among developed countries.[36]

---

[32] https://greendigitalfinancealliance.org/, Green Digital Finance Alliance.

[33] https://www.iisd.org/home, International Institute for Sustainable Development.

[34] http://www.cop21paris.org/, UN Framework Convention on Climate Change 21st Conference of the Parties 2016.

[35] https://www.greenoptimistic.com/european-investment-bank-support-green-investment-20170125/#.WJdP3IWcGms, "European Investment Bank Pledges Unwavering Support for Green Investment," Nicholas Say, *The Green Optimistic*, January 26, 2017.

[36] https://www.huffpost.com/entry/swedish-politicians-troll-trump-administration-while-signing-climate-change-law_n_58948c3be4b0c1284f2558ca?ncid=fcbklnkush pmg00000063, "Swedish Politicians Troll Trump Administration While Signing Climate Change Law," Jenavieve Hatch, *Huffpost*, February 3, 2017.

Update: On June 1, 2017, then-President Trump announced that the U.S. would abandon the Paris Agreement, citing as the reason as it being detrimental to the U.S. economy.[37] As soon as UN rules allowed, the Trump administration formally notified the UN of its withdrawal in November 4, 2019.[38] Due to legal safeguards built into the Agreement, it took until November 4, 2020, for the U.S. to formally leave.[39] This was one day after Election Day in 2020, with Joe Biden eventually certified as the 46th President of the U.S. On January 20, 2021, President Biden reversed the U.S. withdrawal from the Paris Agreement the day he was inaugurated, signaling that the U.S. was intent on resuming international leadership to address climate change.

Deb Haaland Confirmation Fight Over Fossil Fuels

The intentions of two GOP Senators, Steve Daines (R-Montana) and John Barrasso (R-Wyoming), to oppose President Biden's nominee, Deb Haaland, to head the U.S. Department of Interior were reported in the *Huffington Post* on February 8, 2021. Haaland, an experienced and respected Native American Department administrator, is perhaps uniquely qualified to be Secretary of the Interior.

GOP opposition to her presents many stark examples of how deeply divided America is today.[40] The splits encompass all of the following divides:

---

[37] https://www.npr.org/2017/06/01/531090243/trumps-speech-on-paris-climate-agreement-withdrawal-annotated "Trump's Speech On Paris Climate Agreement Withdrawal, Annotated," NPR, June 1, 2017.

[38] https://www.nytimes.com/2019/11/04/climate/trump-paris-agreement-climate.html "Trump serves notice to quit Paris Climate Agreement," Lisa Friedman, *The New York Times*, November 4, 2020.

[39] https://www.bbc.com/news/science-environment-54797743 "Climate change: US formally withdraws from Paris agreement," Matt McGrath, *BBC News*, November 4, 2020.

[40] https://www.huffpost.com/entry/steve-daines-deb-haaland-interior-department-nominee-biden_n_6021b549c5b6c56a89a3b45d "GOP Senators Backed an Extremist at Interior but Reject a Native American Woman," Chris D'Angelo, *Huffpost*, February 9, 2021.

- Exploitation of natural resources versus protection of them
- The economy versus the environment
- Fossil fuels versus clean energy
- Climate denialism versus climate concern
- Stasis versus progress
- Exclusion versus inclusion
- Misogyny versus gender neutrality
- Privilege versus equality.

The GOP Senators opposing Haaland's confirmation are not just short-sighted. If they succeed in blocking her confirmation, they condemn us and our children to a future none of us would want to endure.

Update: Deb Haaland was confirmed as Secretary of the Interior by the U.S. Senate on March 15, 2021, by a vote of 51 to 40.[41]

Protect U.S. Labor Rights, Protect Our Climate

Dharna Noor, writing for earther.gizmodo.com, makes a compelling case of why a piece of labor legislation that never mentions the weather or the climate is nevertheless a key component of the Green New Deal.[42] Noor sees the Green New Deal as a desirable end in itself; it can also be a means to fulfilling America's promise and preserving its democracy and the world's climate.

The Protecting the Right to Organize Act (PRO)[43] is a climate issue, which is partly why passing the Democrats' PRO bill is so important. It will facilitate America's transition to a clean energy economy by protecting existing workers displaced from fossil fuel jobs in coal mining, oil

---

[41] https://www.npr.org/2021/03/15/977558590/deb-haaland-confirmed-as-first-native-american-interior-secretary "Deb Haaland Confirmed as 1st Native American Interior Secretary," Nathan Rott, *NPR*, March 15, 2021.

[42] https://earther.gizmodo.com/why-the-pro-act-is-part-of-a-green-new-deal-1846441751 "Why the PRO Act Is Part of a Green New Deal," Dharna Noor, *Gizmodo*, March 10, 2021.

[43] https://www.congress.gov/bill/117th-congress/house-bill/842 H.R.842 — 117th Congress (2021–2022): Protecting the Right to Organize Act of 2021. (2021, March 11). https://www.congress.gov/bill/117th-congress/house-bill/842

production, and natural gas fracking, and paying them as they retrain for clean energy jobs. As several studies have concluded, it will also create millions of new jobs in a green economy. In one area directly related to climate change, many workers will be needed to cap the thousands of abandoned oil wells that dot the country, many of which are leaking methane, a potent greenhouse gas.[44]

One climate-related area of job growth is the construction of thousands of new energy-efficient homes. Another vital new job sector is electrifying the millions of houses, offices, and factories that use gas for heating, cooling, and cooking. Part of that task is ensuring that the source of electricity is renewable, thereby reducing greenhouse gas emissions from fossil fuel. Just in New York City, the task is monumental, but if the country is to achieve net-zero emissions by 2050, the effort must be nationwide. The Heating, Ventilation, and Cooling (HVAC) sector will boom.

Front and Center: President Biden's Climate Plan

President Biden is putting climate change front and center in his administration. During the 2020 presidential campaign, candidate Joe Biden proposed a less ambitious four-year, $2 trillion climate plan.[45] After his election, President Biden formed six teams to develop policy positions and proposals. He named Alexandria Ocasio-Cortez, Bernie Sanders, and others to the climate team, which developed a much more robust and far-ranging proposal than Biden's campaign one. Biden accepted most of the team's proposal recommendations, announcing a 10-year, $8 trillion plan during his first week in office.

Biden's climate plan is ambitious and comprehensive. His vision is nothing less than a total transformation of the American energy economy from one dependent on fossil fuels — coal, oil, gasoline, and natural

---

[44] https://www.reuters.com/article/us-usa-drilling-abandoned-specialreport/special-report-millions-of-abandoned-oil-wells-are-leaking-methane-a-climate-menace-idUSK-BN23N1NL "Special Report: Millions of abandoned oil wells are leaking methane, a climate menace," Nichola Groom, *Reuters*, June 16, 2020.
[45] https://www.bbc.com/news/world-us-canada-53411506 "Biden sets out $2tn plan for carbon-free electricity by 2035," *BBC News*, July 14, 2020.

gas — to an economy that relies 100% on clean energy sources, principally the sun, wind, and water.

Biden's plan calls for achieving a net zero world by 2050.[46] "Net zero" is a world where any heat-trapping, climate-warming greenhouse gas emissions are offset by additions to Earth's carbon-absorbing capacity. By achieving net zero, we will build a stronger, more resilient nation capable of responding to the planetary challenge of climate change.

Biden brings equity issues directly into the Climate Plan, promising to promote environmental justice for communities of color and low-income ones that polluters have harmed more than affluent, majority-white ones. He also vows to ensure that the workers and communities who have built our economy over decades will not be left behind by the transition to a clean energy economy and that jobs in the new "net zero" climate economy will be good jobs.

Biden also wants to rally the rest of the world to meet the climate change challenge because the threat is global. It respects no boundaries, no frontiers, so it does little good to just clean up our own economy if other countries do not clean up theirs too.[47]

There is much that Biden can do by Executive Order to achieve his ambitious climate agenda, and he has issued several climate-related ones already. Notably, on his first day in office in January 2021, Biden rejoined the Paris Agreement, from which President Trump had announced his intention to withdraw in 2017. That withdrawal process had become final on November 4, the day after Biden's election. So, technically, the U.S. only withdrew from the Paris Agreement for two months.

However, on the legislative side, passage of bills implementing economy-wide climate action is challenging. President Biden is committed to seeking bipartisan support for his climate agenda, but the Senate is much more partisan than it was when Biden served in it, and the country is more divided. The chances of getting 10 Republicans to support a Democrat-proposed Climate Plan bill is slim. Barring that, the Democrats

---

[46] https://apnews.com/article/2ad4e1c11f89436890748a137feff930 "Joe Biden's $5T climate plan: Net zero emissions by 2050," Bill Barrow, *Associated Press*, June 15, 2019.

[47] https://grist.org/politics/joe-biden-executive-order-climate-change-financial-risk-economy/ "Biden's latest executive order takes aim at climate change's risk to the economy," Naveena Sadasivam, *Grist*, May 21, 2021.

can use the reconciliation process to push Climate Plan bills through Congress, which bypasses the need for a super majority in the Senate.[48]

In November 2021, President Joe Biden signed a $1 trillion bipartisan infrastructure bill that includes historic funding to protect the country against the detrimental effects of human-caused climate change. However, the bill did less to curb greenhouse gas emissions. The spin-off Build Back Better Act, which designated $555 billion to aggressively combat climate change by cutting greenhouse gas emissions, was in abeyance for a long period due to lack of full support in the Democratic Party.

President Biden's final recourse is to appeal over the heads of the Senate to the public, in the hope that popular pressure will induce enough senators that it is in their own best interest to support his climate agenda or face voters' wrath come next election. With nearly 50 years' experience in the U.S. government, President Biden is an astute strategist for achieving national and international leadership on climate, but citizen support is essential as well.

Update: U.S. Congress Passes Landmark Climate Legislation

The U.S. Congress finally passed a climate action bill in August 2022.[49] While the legislation is called the Inflation Reduction Act, the preponderance of the bill's provisions relate to GHG emissions reductions. Rather than imposing a carbon cap and trade system, the bill focuses on providing incentives for the production and use of renewable energy in the U.S. and subsidies for electric vehicles so that many more, especially low-income, people can buy them. The stunning aim is to achieve the U.S. target of reducing emissions by 50% below 2005 levels by 2030.

---

[48] https://www.nbcnews.com/politics/congress/what-budget-reconciliation-explainer-fast-track-process-covid-relief-n1256592 "What is budget reconciliation? An explainer on the fast-track process for Covid relief," Sahil Kapur and Frank Thorp V., *NBC News*, February 6, 2021.
[49] https://climateyou.org/2022/08/26/our-take-u-s-congress-passes-landmark-climate-legislation-by-climateyou-guest-contributor/

China, India, Europe, and the Oil Producers

Oil refineries (New Jersey, 2008)

China's Xi Jinping, Marxism, and Climate Change

The piece titled "At the height of his power, China's Xi Jinping moves to embrace Marxism," published by CNN, is an excellent analysis. It is helpful also to look at the dangers and uncertainties Xi faces as he takes China down the state capitalist road with a Marxist gloss.

The greatest challenge facing China is whether it can decarbonize its ever-growing economy fast enough to ensure a livable environment for future generations. The toll of pollution on public health in China is already high. Every coal-fired power plant, old or new, raises the cost in lost productivity, more hospitalizations, and shortened lives. The same is true for every oil-fired plant or factory, and to a somewhat lesser extent, to every plant or factory run on natural gas.

At the United Nations General Assembly in September 2020, China committed itself to reach peak CO_2 emissions before 2030 and to reach carbon neutrality by 2060.[50] World leaders and climate activists welcomed the announcement, but considerable skepticism remains as to whether China is serious about taking the herculean steps needed to reorient its vast, heavily fossil fuel-dependent economy to green energy sources. And, if it is, whether it can actually accomplish such a massive transformation.

It is absolutely essential that China succeed in this endeavor, given that it is the world's largest emitter of greenhouse gases by far. If China does not work hard to green its economy or if it tries but fails badly, the world's inhabitants, both human and wildlife, will have to contend with an evermore hostile climate that will kill many of the old, the very young, the sick, or infirm; it will make life miserable for most of the rest.

Can the Chinese State manage the economy better than the market-driven capitalist system? History would say probably not or certainly not. China is undergoing an intensive migration from the countryside to the cities. Will a managed economy be able to provide employment for the millions who have fled the land that no longer sustains them? In an era

---

[50] https://www.bbc.co.uk/news/science-environment-54256826 "Climate change: China aims for 'carbon neutrality by 2060,'" Matt McGrath, *BBC News*, September 22, 2020.

increasingly automated, computerized, digitized, and robotized, the answer is very probably not.

Will China adopt a universal basic income (UBI)? Unlikely, but it may have to. China's economy also faces a huge demographic challenge brought on by decades of its one-child policy, namely a rapidly aging society with insufficient children to care and provide for elderly parents, and an inadequate or non-existent social safety net. China's 1.4 billion people are its greatest asset and its greatest liability.

Politically, Xi must prove wrong the maxim that power corrupts and absolute power corrupts absolutely. Xi himself may be incorruptible, but what about provincial officials, local factory owners, and billionaire tycoons?

Socially, Xi must incorporate China's ethnic minorities into his vision, not alienate them or suppress their aspirations. He must somehow handle the (inevitable?) revolution of rising expectations. As the masses get some, would they not want more and more? The chasms that have developed in China between the very rich and the no longer desperately destitute are huge and must be negotiated. And when mere survival is no longer one's only concern, will demands not arise for control of other aspects of life, for the freedom to voice concerns or demands, to dissent, to oppose, or to disrupt?

The powers of the State to surveil its citizens with ubiquitous video cameras and facial recognition software integrated with huge databases has increased. Through pervasive bureaucratic means, its ability to control the movement and place of residence of its peoples has reached unprecedented levels. But State control is and will remain tenuous. Can the Party rule the masses? For a long time, yes; but forever, no.

Xi's espousal of Marxism is a direct challenge to democracy. Can State capitalism infused with some Marxist principles build a vibrant, self-confident, sustainable society in the 21st century? Can it overcome the many challenges facing Chinese society today and in the foreseeable future? Besides the challenge of a rapidly urbanizing society and a concomitant depopulation of many rural areas, China will have to adjust to its changing demographic with a rising proportion of older people. Together with urbanization, this trend threatens traditional family-oriented intergenerational services. Who will care for parents if most or all of the children have moved to the cities? Who will care for the few children there are if both parents are off working? What if because of rapid digitization there are not enough jobs for all who want and need them? Will the Chinese State have to develop a social safety net? Will it be forced to implement some form of Universal Basic Income (UBI)? How will the State feed its people if most of the farmers have had to move to the cities in order to survive?

The challenges confronting China's autocratic state capitalist system of government are many and daunting. Several are existential. In this, China bears an odd resemblance to the situation the U.S. finds itself at start of the decade of the 2020s. Both countries, both systems, must provide a tolerable and at least the hope for a better life for all or at least the great majority of its citizens.

Climate change is an existential issue as much for China as for the U.S., indeed for all humanity. If the two superpowers who are also super-polluters do not curtail greenhouse gas emissions roughly by half in this decade and completely by 2050, neither great nation will retain much of its greatness. Neither will have a comfortable present, much less a hospitable future. The U.S., along with the other Western democracies, must deal with similar existential threats.

While the two major systems of governance may devolve into hostile competition reminiscent of the Cold War between the U.S. and the Soviet Union (and now reviving over the Ukraine), all sides are only too aware that the outcome of a nuclear exchange would be mutual annihilation. Therefore, while the possibility of a miscalculation or a miscommunication that could trigger a nuclear holocaust always exists, a far more likely outcome will be a mutual if uncodified understanding that the two power centers will both compete and collaborate when it is in their interest to do so.

China and the U.S. are "frenemies" writ large. Each power will strive to demonstrate the superiority of their model. The contestants have taken the stage, the rules of the game are being established, the red lines implied or demarcated, the first challenges issued, the first skirmishes soon to be joined. Both players know that not only are their regimes and their countries' standing in the global community at stake, but so too is the fate of the world.

A significant area of competition in the new Cold War is leadership on climate change action, now that President Biden has rejoined the Paris

Agreement. Both superpowers are vying for climate change leadership and the economic benefits that will accrue to the winner in clean energy technology. That competition may not be a bad thing since it will likely accelerate the required technological development. Plus, we need everyone on the planet — living and working under both hegemonies — to transform to carbon neutrality in the coming decades.

Can China and America Stop Climate Change?

David E. Sanger credits U.S. President Joe Biden with uncommon candor in his White House Memo published on March 26, 2021, in the *New York Times*.[51] Rarely does a world leader lay out the guiding challenge of his administration with such clarity as Biden did in his first press conference. Biden's answer — prove that democracy works and that it is better than authoritarian rule.

Beyond the competition between radically different governance systems, areas where China and the U.S. have got to work together are world health (preparing for and trying to prevent the next pandemic), nuclear containment (any nuclear exchange means curtains for all humankind), and most crucially, curbing climate change (failure means a slower but equally certain end to the human experiment).

If only one of the world's two largest emitters of greenhouse gases decarbonizes, both will be plunged into climate catastrophe. Both must come very close to having zero net emissions or chaotic, horrific conditions will threaten to swamp civilization.

Biden thinks he has the cure for what ails America. He has a bold, transformative agenda, but before he can apply the medicines the country needs, he must enact them into law. To do that he must overcome — or subvert — the Senate filibuster rule that lets a minority thwart the will of the majority. The filibuster is an anachronism that lately has been used to

---

[51] https://www.nytimes.com/2021/03/26/us/politics/biden-china-democracy.html "Biden defines his underlying challenge with China: 'Prove Democracy Works,'" David E. Sanger, *The New York Times*, March 26, 2021.

obstruct all legislation proposed by the opposite political party, lest it get credit for it in the next election.

Yet, Americans have got to know that their democracy is not in the best of health these days. In fact, it was close to terminal on January 6, 2021, when the nation's Capitol was assaulted by an insurrectionist mob and remains in critical condition. Can it recover its health, its confidence, and its competence?

It must be said, however, that the relationship of the U.S. with China must entail cooperation as well as competition. Without the props of civilization, humankind will go the way of the dinosaurs unless, led jointly by America and China, it can mobilize its big brains, many talents, great adaptability, enormous wealth, scientific method, formidable technology, innate optimism, and survival instinct to halt the heedless disequilibrium we have caused in the climate.

We can restore the fine balance we need to survive, but only if all nations act urgently, decisively, and effectively. We all have to act as if our lives depend on what we do to counter climate change in this decade because our lives and those of our children and theirs do depend on it. If we do not succeed, neither America's democratic system nor China's autocratic one will be superior to the other. If collectively we do not stop the climate from over heating, the Earth will become unlivable for humans, and both systems of governance — China's and America's — will have failed the only test that really matters, the test of establishing a sustainable future for our planet.

Update: U.S. and China Agree to Joint Action

The relationship between the U.S. and China has been tense lately, but their agreement before President Biden's Climate Ambition Summit on April 22, 2021, to work together on climate change is heartening.[52]

---

[52] https://www.axios.com/us-china-agree-joint-climate-action-838a35b5-92a8-47c6-901f-1a7e9926f5bc.html "U.S. and China agree to take joint climate action," Andrew Freedman, *AXIOS*, April 28, 2021.

An agreement between the U.S. and China to work jointly on curbing the climate crisis is very good news. If the Paris goals are to be met, China and the U.S. must lead the way, curbing their world-largest greenhouse gas emissions and shepherding a global march to control climate change. Secretary of State Blinken's meeting with a Chinese delegation in Anchorage in March did not go well; President Biden's climate envoy John Kerry's encounter with his Chinese counterpart in Shanghai could have likewise achieved little.

Instead, perhaps because Kerry met with Xie Zhenhua, with whom he had worked on the Paris Agreement in 2015, the two sides both agreed to cut carbon emissions in the 2020s, although no figures were specified. Still, given the context of strained Sino–U.S. relations over tariffs, intellectual property, Hong Kong, Uighurs, Taiwan, and the South China Sea, even an agreement in principle to work together on climate is a cause for optimism, if not yet celebration.

India's Huge Carbon Footprint

India is the third highest emitting country, behind only China and the U.S. All three top emitters have to accelerate their pace of decarbonization if the world is to have a reasonable chance for meeting the Paris Agreement targets. In the energy sector, current Prime Minister Narendra Modi has prioritized building up India's solar capacity, but has lagged on closing coal mines and coal-fired power plants. India still imports a lot of oil and, increasingly, liquid natural gas (LNG). LNG is seen as a bridge fuel to lower greenhouse gas emissions, but it still contributes to India's GHG emission total. KN Hari Kumar has authored a trenchant two-part analysis of the similarities and differences between conditions facing Indira Gandhi in 1974 and those that Modi operates in today.[53] Nehru laid out the

---

[53] https://scroll.in/article/882994/nehru-may-have-constrained-the-bjp-from-returning-to-its-blatant-hindutva-agenda-of-the-1980s "1974 again? Despite wielding enormous power, Modi has failed to destroy India's Nehruvian idealism," KN Hari Kumar, *Scroll.in*, June 25, 2018.

vision of a united, democratic, secular, developed India. The Congress Party under Indira Gandhi became the embodiment of that vision.

One of the two major political parties in India along with the Indian National Congress, the Bharatiya Janata Party (BJP), shares some aspects of Nehru's vision, especially that of development, but not others. The BJP is religious and separatist, espousing Hindutva (a Hindu-dominated State). Yet, while Modi has promoted inclusive development, he has been constrained from imposing the harshest pro-Hindu/anti-Muslim measures.

Will economic development proceed fast enough and broadly enough to enable Modi to continue to disappoint the more radical religious elements of his party? Will Modi be able to meld Nehru's vision for India with the BJP's? Or are they essentially incompatible? India held national elections in 2019, which Modi and the BJP won convincingly. He has retained his popularity among the Hindu majority, despite a stumble over the sudden withdrawal of all currency over 1,000 rupees (about $13) in 2016, for the stated reason of curbing hoarding and the black market. Shopkeepers screamed in outrage but adjusted within a few months.

A move with more serious repercussions was a 2019 amendment to the Indian Constitution that treated Muslims as second-class citizens. Not surprisingly, Muslims rioted in Delhi and other major cities. The disturbances have abated, but the resentment lingers. Modi's administration has announced plans for a national database of all Indian citizens. Implementation has been deferred, but it is a move that hangs like a cloud over the Muslim community.

Modi is a populist who got along famously with President Trump. Under Biden, the U.S. will seek to keep India as an ally, and promote trade, development, and more rapid decarbonization, while urging India not to edge toward break-up. India's status on the world stage depends in part on how fast India continues its rapid development that has seen millions lifted out of poverty into the Indian middle class. It also depends on how quickly it achieves its transition from coal-based energy to low-carbon alternatives, and on whether its Paris Agreement climate targets are met.

India now emits 2,200 MT of carbon dioxide a year. At the 2021 COP26 in Glasgow, Modi vowed to increase non-fossil energy capacity to 500 GW (gigawatts) by 2030; meet 50 percent of energy requirements from renewable energy by 2030; reduce India's total projected carbon emissions by 1 billion tonnes by 2030; and achieve net zero carbon by 2070.

Feeding India's Hunger for Oil

Oil conveys power and profits, so it is in the interests of the United States, Saudi Arabia, Israel, and the United Arab Emirates (UAE) to align quietly to sell India the oil it needs and wants as a counterweight to a China seeking markets and influence in Asia. What is striking about the informative article by Simon Watkins, titled "How India's Hunger for Oil Could Transform the Middle East,"[54] and published February 15, 2021, on oilprice.com, is that everyone, author and subject nations alike, assumes that the international power game continues, and will continue, to be played as it has for almost a century. Oil is the lubricant, the essential elixir, the life blood, the *sine qua non* of power and prestige.

That assumption no longer holds. Much of the world has awakened to the negative consequences of our dependence on "black gold" for everything from our daily survival to our enduring civilization. Combustion of oil and other fossil fuels emits heat-trapping gases, the so-called greenhouse gases (carbon dioxide, methane, and nitrous oxide) that are heating the climate and causing it to turn hostile to human life. Catastrophic conditions loom if greenhouse gas emissions are not cut drastically by 2050. Decarbonization of the world's energy system is a must if we are to avoid any of several nightmare scenarios.

India's 1.4 billion people can rejoice that its government wants to better their lives. Switching to oil-generated power from today's prevalent coal-fired plants is an improvement, but a power plant built today will spew emissions for 40 years. By the time that plant is retired, the climate

---

[54] https://oilprice.com/Energy/Crude-Oil/How-Indias-Hunger-For-Oil-Could-Transform-The-Middle-East.html "How India's Hunger For Oil Could Transform The Middle East," Simon Watkins, *OilPrice.com*, February 15, 2021.

will be killing people. Indians would be somewhat better off if India were to opt for natural gas instead of oil because gas is cleaner than oil, but best of all would be if India were to develop its economy and society almost exclusively through renewable energy sources such as wind, sun, and water. Because India's economy is so large and growing so rapidly, the path it chooses is critical to humanity's future. Only China's energy path will be as determinative.

For the U.S., the UAE, Kuwait, and others to addict India to oil in order to counterbalance China's assertive drive for influence in the Middle East is worse than folly. It is to fast-track the human race to a hellish existence of intolerable heat waves; years-long crop-killing droughts; cities flooded every full moon; frequent devastating tropical storms; lengthy, ever more destructive wildfire seasons; frigid winters caused by climate-disturbed polar vortexes; hunger-caused political unrest leading to regime collapse; and climate refugees numbering more than 100 million per year. Such an existence is best avoided. It can be, but only if we begin curbing greenhouse gas emissions now in order to reach net zero emissions by 2050.

Getting to Net Zero — Can the UK Do It?

A good article entitled "What would life be like in a zero-carbon country?" by Ivan Kottasova, writing for CNN, details what getting to net zero carbon emissions would mean for the United Kingdom's citizens.[55] "Net zero" means the amount of greenhouse gases emitted into the atmosphere is no more than the amount taken out. Except for Brexit's damper on the UK's economy, Americans can expect similar effects. Basically, the conclusion is that Britons will be able to deal with the transformation, but that some industries will find it difficult to make the transition to zero carbon emissions by 2050. The most challenged industries include agriculture, aviation, and shipping.

Some say the cost, perhaps £1 trillion ($1.4 trillion), of converting to net zero is too high; others warn that the risks and costs of the status quo

---

[55] https://www.cnn.com/2019/06/16/uk/net-zero-emission-target-gbr-intl/ "What would life be like in a zero-carbon country?" Ivana Kottasová, *CNN*, June 16, 2019.

far exceed the costs of transition. Some argue that implementation should be faster; more caution that faster is not doable.

Can the UK government deliver on its net zero promise, given the long and difficult process it underwent to finally deliver on Brexit? The Boris Johnson government had the support of the people for cutting emissions and faced no major-party opposition on this issue in the Parliament. The UK took its hosting of the 26th Conference of the Parties of the United Nations Framework Convention on Climate Change very seriously. It endorsed the European Union's unequivocal support for a green rebuilding for the COVID-stricken economy. However, the UK was consumed by efforts to gain the upper hand over the coronavirus pandemic, with tussles over the pace of vaccination and the extent and duration of lockdowns. The government's capacity, its commitment, and its staying power to deliver on its promise of a Green UK are all in question. Time will tell if the UK is able to achieve its ambitious target of net zero emissions by 2050.

France Judged Guilty in Climate Case

Climate activists have touted the 2015 Paris Agreement as a milestone in the commitment made by almost all the countries of the world to limit climate change. Recently, however, there is a Paris judgement against the French state for its failure to address the climate crisis.[56] This case has been called the *L'Affaire du Siecle* (The Case of the Century). It is the first time that a court has held a country accountable for inaction to preserve the future for its citizens. The precedent has been set, and other judgements will follow. The first step has been taken; the first domino has fallen, and climate activists everywhere cannot rest.

The case was initiated by four non-governmental organizations (NGOs) beginning in late-2018, citing governmental liability for

---

[56] https://www.theguardian.com/environment/2021/feb/03/court-convicts-french-state-for-failure-to-address-climate-crisis " Court convicts French state for failure to address climate crisis," Kim Wilsher, *The Guardian*, February 3, 2021.

ecological damage caused by climate change.[57] The four NGOS that brought the suit are the *Fondation pour la Nature et l'Homme* (FNH), Greenpeace France, *Notre Affaire a Tous*, and Oxfam France. In France, 2.3 million people have signed a petition named *L'Affaire de Siècle*[58] that calls on the French Government to respond much more effectively to climate change.

The NGOs challenge the French Government's inaction on climate change and its failure to meet its stated goals to reduce greenhouse gas emissions, increase renewable energy, and limit energy consumption. They cite the general principle of law that provides for the right of everyone to live in a "preserved climate system" and "a healthy and ecologically balanced environment." They argued that the French government has a "duty of care" to "take all necessary measures to identify, avoid, reduce and compensate the consequences of climate change." These measures include implementation of a legislative and regulatory framework and adoption of practical measures meant to fight efficiently against climate change.

The Administrative Court of Paris found that prevention of ecological damage is included in the Civil Code of France. It based its judgment in part on reports from the Intergovernmental Panel on Climate Change (IPCC), the United Nations Framework Convention on Climate Change (UNFCCC), the Paris Agreement, European climate directives and regulations, the *Charte de l'Environnement* (French Environmental Charter with constitutional standing), and the Energy Code. The latter sets national greenhouse gas emissions targets and an accompanying low-carbon national strategy. The targets set were a 40% greenhouse gas emission reduction in 2030 and carbon neutrality in 2050.

From these documents, the Court inferred that the Government of France recognizes the urgency of the climate crisis, has made international commitments to limit the rise of global temperatures caused by

---

[57] http://climatecasechart.com/climate-change-litigation/non-us-case/notre-affaire-a-tous-and-others-v-france/ Notre Affaire à Tous and Others v. France (2018).

[58] https://laffairedusiècle.net/petition, L'Affaire de Siècle Petition.

greenhouse gas emissions, and committed to use its regulatory powers to reduce emissions at the national level.

The Court then evaluated how the Government of France was doing in regard to its adopted greenhouse gas goals. It concluded that France's targets were not being met and ordered the government to comply with its own targets. The Administrative Court of Paris recognized that by failing to act, the French government has caused ecological damage from climate change.[59]

There is a dramatic increase in climate lawsuits in many parts of the world.[60] The United Nations Environment Programme (UNEP) reports that as of July 1, 2020, at least 1,550 climate change cases had been filed in 38 countries.[61] Two cases similar to the French case have been filed in the Netherlands and Ireland. These cases against governments who are slow in responding to their Paris Agreement commitments are a powerful tool for igniting action on climate change. The UNEP report's authors argue that the striking increase in climate cases around the world will be a driving force in bringing action on climate change.

What Happens if the U.S. Quits the Middle East?

Simon Tisdall speculates in his article titled "Why instinct and ideology tell Trump to get out of the Middle East," on what would happen after the closing of U.S. military bases in Iraq, Saudi Arabia, Bahrain, Kuwait, Qatar, United Arab Emirates, and Oman.[62] While there would be both good and bad effects, Tisdall was generally in favor of an American exit. He argued

---

[59] https://www.vox.com/2021/2/4/22265316/france-climate-change-paris-court, "A court has convicted the French government of failing to meet its climate goals," Jariel Arvin, *Vox*, February 4, 2021.

[60] https://climate.law.columbia.edu/, Sabin Center for Climate Change Law, Columbia Climate School, Columbia Law School, Columbia University in the city of New York.

[61] https://wedocs.unep.org, United Nations Environment Programme Document Repository.

[62] https://www.theguardian.com/us-news/2020/jan/11/why-instinct-and-ideology-tell-trump-to-get-out-of-the-middle-east-suleimani-iran "Why instinct and ideology tell Trump to get out of the Middle East," Simon Tisdall, *The Guardian*, January 11, 2020.

that getting out would probably be, on balance, good for the U.S., although the country would have to cope with the diminution of global status and influence. A departure on this scale could be a difficult and long-term convalescence, as the Brits are still learning.

How is the U.S. withdrawal from the Middle East a climate story? The region's power comes from its huge energy resources of oil and gas. But fossil fuels, we have learned, have a downside. They have powered our civilization, enabled our growth and development, and conferred upon us comfort, enough food to eat, and long lives ever since the Industrial Revolution began. The cost, however, has been to so alter the atmosphere with the heat-trapping gases released by combustion of coal, oil, and natural gas, that the climate is heating more rapidly than it ever has in the 4.5 billion-year-long history of the planet.

This rise in global temperature has cascaded through the complex interlocking set of climate subsystems to effect changes that threaten to make our lives miserable, if not un-survivable. Therefore, if the human race is to avoid the severest consequences of global heating, it must stop powering its comforts by burning fossil fuels. Until recently, it has gotten most of its oil and gas from the Middle East, a fact that gave the region immense wealth and influence. The world has awakened to the negative aspects of energizing its civilization with fossil fuels. It is transitioning from carbon-based fuels to sun-, wind-, and water-based energy sources, resulting in the Middle East losing importance and facing upheaval as its income stream evaporates and its own climate worsens.

These circumstances enable the U.S. to reduce its military presence in the Middle East, which will have mixed but slightly positive results both for the region and the U.S. Whether and how the U.S. re-engages in the future depends to some extent on events there, but more so to events in Washington. The next decade or two in the Middle East will be dominated by local and international reactions to fallouts from the climate-caused energy transition. They will be turbulent times that the U.S. can help ease through financial, technological, and humanitarian assistance to Middle Eastern nations. If the U.S. does not do it, China surely will.

Tisdall is optimistic, perhaps overly so, that our leaving would be good for the Middle East. Iran clearly has designs on becoming the next hegemon in the region. Without American resistance, those designs could proceed apace, resisted only by Sunni states for whom Iranian Shia domination is anathema. ISIS — the Islamic State of Iraq and Syria — would have more scope to rebuild and renew its terrorist attacks.

The ultimate question is how vital to American interests is the Middle East? Today's answer, and tomorrow's, is "not very."[63] We are no longer dependent on the region for oil, thanks to the fracking boom that made the U.S. energy independent, but fracking is itself already facing decline and, although unthinkable to oil producers, obsolescence. With ever-diminishing oil revenues and ever-increasing pressures from heat waves, droughts, and attendant crop failures, the oil-dependent countries of the Middle East will be hard-pressed to maintain their standards of living and the loyalty of their subjects. Some will adapt and survive, others will not. The region will be turbulent and impoverished.

Benign neglect by the U.S. at that stage would be a mistake; re-engagement could promote less autocratic regimes in at least a few Middle Eastern countries. So, the wisest course for America now is to withdraw militarily while retaining diplomatic ties; not to forget about the Middle East, and not to ignore it. It should exert soft power when and where appropriate, uphold American values, support democratic movements, champion human rights, facilitate economic and social development, and engage with the region on climate policy that provides a transition away from the region's lifeblood — oil.

Finally, U.S. policy should focus on addressing our own climate challenges and democratic shortcomings. The U.S. needs finally to undertake serious climate action and long-overdue reforms, tackle the long-festering

---

[63] This assessment may be changing due to the war in Ukraine and the desire to reduce dependence on fossil fuel sources from Russia. President Biden's rekindling of relations with Saudi Arabia to secure oil supplies in 2022 is a case in point.

failures in both climate action and democracy, and achieve its much-touted ideals of global climate leadership, as well as liberty and justice for all.

The Fate of the Oil Producers

The global energy transition needs to happen fast, but the consequences for countries dependent on oil exports for much of their revenue could be catastrophic. Big rich countries like Saudi Arabia can diversify. For example, the Saudis are building a plant to make green hydrogen. Russia too has a large, diversified economy, but the oil sanctions imposed by European countries due to Russia's invasion of Ukraine in 2022 are taking a toll on its economy. Norway has a huge sovereign fund. Iraq, however, tops the list of countries that derive most of their revenue from the export of oil and gas, getting 90% of its Gross Domestic Product (GDP) from oil exports. Other countries facing steep declines in national budgets include Libya, Venezuela, Equatorial Guinea, Nigeria, Iran, Guyana, Algeria, Azerbaijan, and Kazakhstan.[64]

Government services, subsidies, and employment in many countries will have to be curtailed. That constriction of affected countries' economies bodes poorly for popular unrest as cuts hit home. Regimes that can well manage the shrinkage and the turmoil it creates will survive. Those that do not, would not. Venezuela is a harbinger of what to expect: shortages of many goods in shops and markets, including food staples and most imported items, an increasingly desperate citizenry, and burgeoning emigration.

Now imagine the turmoil that multiple Venezuelas would cause. There are 40 countries that export over $1 billion in oil and gas annually. Some will be able to handle cutting that revenue in half within a decade; many would not. As the energy transition to renewable fuels gains momentum, demand for oil and gas will fall, and so will the price of oil, further eroding

---

[64] https://oilprice.com/Energy/Crude-Oil/Rapid-Energy-Transition-Could-Doom-Oil-Exporting-Countries.html "Rapid Energy Transition Could Doom Oil Exporting Countries," Alex Kimani, *OilPrices.com*, June 1, 2021.

revenues. The value of proven reserves will also decline, further undercutting the borrowing power of oil-producing countries and companies.

Bankrupt countries do not disappear, but their citizens face deprivation and misery. The Paris Agreement called for members to contribute $100 billion a year to help poor countries make the break from fossil fuels to clean energy. Donations to the fund have fallen way short, while the realization has grown that the cost of the transition will be far higher than anticipated.

Will the Group of Seven (G7) — an informal group of wealthy democracies comprised of Canada, France, Germany, Italy, Japan, the UK, and the U.S. that meet annually to plan coordinated action on major global issues — step up? To some extent, yes, but the rich countries face the task of transitioning their own energy systems, so they will be constrained in providing the vast sums needed for other countries as well.

For rich oil-importing countries, as the energy transition away from fossil fuels proceeds, energy expenditures will decrease, freeing up capital to speed the switch and to support oil exporters. Debt levels of the oil-exporting countries are likely to exceed all previous records, leading to both inflation and defaults. Modern monetary theory — which asserts that because sovereign countries can print money, debt does not matter as long as the money is spent productively — will get a thorough test of its validity.

This decade and the two following decades will likely be highly volatile across the world, with stresses that will roil social, political, financial, and economic realms. There will be serious ramifications for the stability of the international order. The viability of our civilization is based on almost 200 sovereign nations with over a fifth of them oil-exporters. It relies on an economic system, capitalism writ large, that pursues continual short-term growth through the exploitation of natural resources — animal, vegetable, and mineral.

However, we are learning that those resources are finite and that we cannot continue to exploit them as if they were infinite, if we are to preserve a habitable planetary home.

If we humans are to survive, both sovereignty and capitalism must adapt. Nation states must cede some degree of sovereignty in recognition that everyone inhabits the same habitat, where all boundaries are artificial.

Divisions, while important, pale when compared to our commonality: we are all humans; we are all Earth people.

Capitalism must accommodate limits to growth and evolve into a sustainable circular economy. This change will require that we develop an understanding of the interdependency of all life and a new reverence for all forms of life that make this planet home. We must all live together in the world we share, or many forms of life will eventually become extinct.

Can we do it? Can we use our big brains to adapt to new realities, or will we be held captive by old, outmoded ways of thinking, acting, and being that condemn us to repeat past mistakes and follies leading us only to our own demise? The energy transition from fossil fuels to carbon-free, limitlessly renewable energy sources is a test, a big one, of our ability to adapt to new circumstances and challenges. However, it is only the first of many that we must pass if we are to resolve the climate crisis and evolve a civilization that will endure and flourish not for years, decades, or centuries, but for millennia.

Heroes

Mad River Valley at Knoll Farm (Vermont, 2011)

Al Gore at 73: Climate Statesman

"The crisis has continued to worsen much faster than we have been implementing solutions," said Al Gore, near the start of a wide-ranging interview he gave on Earth Day in 2021. The climate crisis is "an existential threat to the future of human civilization." He was speaking with journalist Jonathan Capehart, as part of the online video series, *Washington Post Live*: "The Path Forward."[65]

At 73, Gore has become an elder statesman of environmental advocacy. Over the years, he has written a dozen thought-provoking books, beginning with *Earth in the Balance: Forging a New Common Purpose* (1992; revised 2013). Among many other honors, he has won the Academy Award for Best Documentary for *An Inconvenient Truth* in 2007, also made into an influential book. With the Intergovernmental Panel on Climate Change (IPCC), he won the Nobel Peace Prize that same year, for helping to educate people worldwide about the increasing dangers of a warming planet.

Few in the environmental fight have Al Gore's breadth of experience. As a politician, he served four terms in the House of Representatives in the 1970s and 1980s, two terms in the Senate in the 1980s and 1990s, and two terms as the 45th Vice President of the United States from 1993 to 2001.[66]

After his unsuccessful Presidential bid in 2000, he left the world of elected politics and pursued experiences in different realms: in business, technology, and media; in academia as a professor; and as an executive and board member in non-profit organizations. He co-founded Generation Investment Management and is Senior Partner at the venture capital firm Kleiner Perkins, in the area of climate change solutions. He has an advisory role with Google and sits on the board at Apple, Inc., building ties with the tech world. From 2005 to 2013, he co-owned Current TV network, before it was sold to Al Jazeera America.

---

[65] *Washington Post Live* (2021, April 23), Former vice president Al Gore on the global climate crisis [Video], *YouTube*. https://www.youtube.com/watch?v=Jo2jPZFZ9-o

[66] https://www.nytimes.com/2012/08/26/fashion/the-end-of-the-line.html, "The End of the Line," Patrick Healy, *New York Times*, August 25, 2012.

He travels, consults, and lectures frequently, going wherever he feels "he can make a difference," as a friend told the *New York Times* in 2017. And he proudly leads the Climate Reality Project, which gives training workshops to climate activists in the U.S. and around the world. These varied experiences, and his long, studious absorption in the subject of global warming, have given Gore credible insights into the many facets of climate change.

As expressed to Jonathan Capehart, Gore's view that the crisis is worsening "much faster than we have been implementing solutions" is spoken with knowledge and authority. He points out that the climate change predictions made back in the 1970s have either come true or underestimated the pace of the unfolding crisis.

Hearing Gore speak along those lines might at first give the idea that he is discouraged. Or perhaps fatigued, after a lifetime of ceaseless but so far fruitless efforts to bring the situation under control and galvanize sufficient response.

But in the bulk of his comments, he remains resolutely optimistic. After a quick, early mention of our collective spate of alarming dangers — hurricanes, wildfires, record-breaking heat, the Texas freeze, warming oceans, droughts, and species extinctions — he then moved to the many encouraging signs he sees. It is on the good news that he seemed to prefer to focus.

First, President Biden. Biden's Climate Summit was a success, Gore believes, helping "to rally the nation and rally the world." It signaled that America has once again assumed its proper leadership role.

Then, the cost of renewable energy has come down dramatically. With the new affordability of solar and wind power, and the development of better batteries, Gore called the target to cut greenhouse emissions by 50% by 2030 as "a real goal, and it can be achieved." For example, electric cars are becoming affordable. Plans for new coal plants are being cancelled. Existing plants that burn fossil fuels are being retired. And China, which had announced the construction of many new coal plants, may be on the verge of reconsidering. A quicker move to solar and wind would save China a predicted $1.6 trillion over the next 20 years.

Gore spoke with understandable enthusiasm about a coalition named Climate TRACE. (TRACE stands for Tracking Real Time Atmospheric

Carbon Emissions.)[67] This remarkable initiative uses advanced technology, satellites, artificial intelligence (AI), and other techniques to monitor greenhouse gas (GHG) emissions with great speed, transparency, and detail. Gore said that negotiators and policy planners who are talking about their carbon commitments will not have to rely "on self-reported emissions that are often a year or two out of date, often inaccurate." Instead, "we are going to have real time, highly accurate emissions data from every single country and every source within every country."

Continuing the positive news, Gore spoke passionately about the "sustainability revolution" in the investment marketplace. Fund managers and others have put money into sustainable green companies, while turning more away from fossil fuel companies, especially coal. He said Environmental Social Governance (ESG) investments are outperforming non-ESG investments. Brushing aside the anomaly of some big banks that are "lagging behind, still chasing profits from fossil fuels," he said the move towards green companies "is the most significant investing and business opportunity in the entire history of the world."

In another positive area, he discerns the environmental justice movement gaining momentum, aligned with the Black Lives Matter movement and an important component of that larger effort to combat systemic racism. In his travels, he has noted residents' growing pushback to the practice of locating toxic facilities and hazardous waste dumps "just upwind or upstream from communities of color." As he put it, black citizens and their white allies are "standing up to say no more of this environmental racism."

Finally, Gore cited with satisfaction a growing number and variety of voices making themselves heard on the climate issue. Not only are grassroots activists attending Climate Reality training workshops, but young people around the world are mobilizing and marching. Average voters are demanding change of their elected legislators. The whole world appeared to Gore to be "crossing the political tipping point on climate, right now, right this second."

Ironically, much of this progress is triggered by the worsening crisis itself, as climate change becomes easier to see and feel, less deniable and

---

[67] https://www.climatetrace.org/, Climate Trace.

ignorable. Is it Gore's temperament and good fortune in life that make him see the glass half full? A strategic decision to emphasize hope? Or is it the judgment of a seasoned and dedicated advocate who has sighted a constellation of exciting, encouraging signs at last, along the path of his life's mission?

Greta Thunberg: Youth Activist for Climate

In a video released May 2021 titled "Our Relationship to Nature is Broken,"[68] Greta Thunberg, a vegan, extends her climate change message to focus on food production. At present, she says, 83% of agricultural land is devoted to the raising of livestock, while animals provide only 18% of our food calories. The raising of animals for food leads to deforestation, the destruction of wild habitats, mass extinction of species, the increase in new diseases, and the addition of billions of tons of carbon to the atmosphere. Using the work of scientists as her guide, she "connects the dots," as she puts it, showing climate change, farming methods, and people's consumption of meat and dairy to be interlinked. Her point is that to slow global warming, not only are new political and economic systems needed, but our whole destructive relationship to nature must change.

When we think of Greta Thunberg, she often seems a small, solemn, solitary figure, alone even in a crowd. Alone in front of Swedish Parliament with her homemade sign, "Skolstrejk för Klimatet," or "School Strike for Climate."[69] Alone at podiums or microphones, such as at the United Nations Framework Convention on Climate Change (UNFCCC) Conference of the Parties in Poland[70] and the World Economic Forum in Davos,[71] Switzerland. Alone in the spotlight delivering her extraordinary

---

[68] https://www.facebook.com/watch/?v=227377262111626 Greta Thunberg, #ForNature, Facebook video, May 22, 2021.

[69] https://www.bbc.com/news/world-europe-49918719 "Greta Thunberg: Who is she and what does she want?", BBC News, February 28, 2020.

[70] Connect4Climate, (2018, December 16) Greta Thunberg full speech at UN Climate Change COP24 Conference. [Video]. YouTube. https://www.youtube.com/watch?v=VFkQSGyeCWg.

[71] https://www.weforum.org/agenda/2021/01/greta-thunberg-message-to-the-davos-agenda/ "Greta Thunberg's message to world leaders at #DavosAgenda," Greta Thunberg, World Economic Forum, January 25, 2021.

TED talk.[72] And, alone in her informational videos that she posts to extend the reach of her message.

If she seems like a person separate unto herself, she meanwhile directs her urgent message to everyone in the entire world, from the most powerful leaders to young people like herself, many too young to vote. That message is to treat the present climate crisis like the crisis it is. Curb emissions of greenhouse gases, which are still increasing. Leave the oil, coal, and gas in the ground. Accelerate the switch to renewable energy sources. Limit global warming to 1.5° Celsius (2.7° Fahrenheit) over pre-industrial levels. Raise your awareness and educate yourself. Listen to the climate scientists. Mobilize for climate justice and equity. Save the future, for future generations. Demand action; do not settle for more talk. Be change agents for new attitudes, new political and economic systems, and new habits, choices, and ways of thinking.

Why? Because the clock ticks steadily towards the point of no return, towards the collapse of civilization. Time is running out, and, in her famous words, "Our house is on fire."

In the face of these enormous, daunting issues, Greta Tintin Eleonora Ernman Thunberg, who turned 19 in January 2022, keeps her composure and maturity, her almost otherworldly quality. To her, this independence and self-possession are essential. In a speech she said, "Many people love to spread rumors saying that I have people 'behind me' or that I'm being 'paid' or 'used' to do what I'm doing. But there is no one 'behind' me except for myself." While she has sometimes, she says "supported and cooperated with several NGOs (non-governmental organizations) that work with the climate and environment," the fact is, "I am absolutely independent and I only represent myself."

On many occasions, Thunberg has spoken frankly about her cognitive differences and diagnoses — her neurodiversity. The conditions that she has been found to have — Asperger's, obsessive compulsive disorder (OCD), and selective mutism — no doubt play a role in her determined

---

[72] https://www.ted.com/talks/greta_thunberg_the_disarming_case_to_act_right_now_on_climate_change "The disarming case to act right now on climate change," Greta Thunberg, TED, November 2018.

mission.[73] People with Asperger's syndrome, on the autism spectrum, are often indifferent to social codes, popularity, and social games, and they tend to see things in black and white. Shrugging off the criticisms and bullying of detractors, she makes no apology for seeing things in a stark, bleak way, saying that the looming environmental catastrophe warrants it.

Suffering serious depression in a time before she began to speak out against climate change, Thunberg has said, she had "no energy, no friends, and I didn't speak to anyone. I just sat alone at home, with an eating disorder." But with the help and support of people who cared about her, she told *The View* in 2021, she was able to put her autism to good purpose — to make it into a "superpower."[74]

As an example of this power, she said she is able to focus and study for many hours, concentrating with great absorption. In the past three years, she has taken what she learned about climate change and dedicated herself to speaking out. Her goal, she explains, is to do what is right and to make a difference. She is often asked, "What can we do?" and she invariably replies, "Inform yourself."

Happily, Greta Thunberg is, of course, not alone. She is not alone in at least four important ways.

First, despite their initial skepticism, her devoted parents are now fully supportive of her. The family, including Greta's sister Beata, has written a book together, titled *Scenes from the Heart*, about their lives. Thunberg's father, Svante Thunberg, accompanied Greta to the U.S. and back in 2019 — an arduous trip involving two trans-Atlantic crossings in zero-emissions sail boats. Her mother, Malena Ernman, an opera singer, now takes only local singing engagements that she can reach by train. And Thunberg's parents are supportive monetarily. When Thunberg travels in her role as climate activist, she stresses that she takes no remuneration, adding, "My parents pay for tickets and accommodation."

---

[73] https://www.theguardian.com/environment/2019/sep/02/greta-thunberg-responds-to-aspergers-critics-its-a-superpower, "Greta Thunberg responds to Asperger's critics: 'It's a superpower,'" Alison Rourke, *The Guardian*, September 2, 2019.

[74] https://www.facebook.com/watch/?v=159696079460006 "Greta Thunberg on turning Asperger's into her superpower," *The View*, Facebook video, May 25, 2021.

Nor is Thunberg alone in her role as youthful leader working to bring about change — far from it. *The Independent* has described, for example, the achievements of Licypriya Kangujam, 9, of India; Lesein Mutunkei, 17, of Kenya; Ella and Caitlin McEwan, 10 and 8, of the UK; Lilly Platt, 13, of the Netherlands; Xiuhtezcatl Martinez, 21, of the U.S., and Luisa Neubauer, 25, of Germany.[75]

Similarly, in his article titled "The Left Turn" published in the *New Yorker* on May 31, 2021,[76] Andrew Marantz talks to young data-driven activists, many in their 20s and early-30s, such as Alexandra Rojas, Max Berger, Rhiana Gunn-Wright, Sean McElwee, Waleed Shahid, Varshini Prakash, Guido Girgenti, Evan Weber, and Yong Jung Cho. One of their central agenda items is to "decarbonize the American economy" and ensure "a livable planet," via a Green New Deal. As Marantz points out, "Bringing about this kind of fundamental political change is not easy work for anyone, much less a small cadre of near-neophytes." Still, it is clear that many capable, dedicated young people have emerged to do the hard work in their areas, as Thunberg has in hers.

There is another group that Greta Thunberg is not apart from, but a part of — the millions she reaches with her message. Thunberg has, at this time, 11.4 million followers on Instagram, 4.9 million on Twitter, and a vast Facebook audience. Hundreds of thousands of young people all over the world have held school strikes and "Fridays for Future" marches, following the pattern of her initial solo protest. Her slim but powerful book titled *No One Is Too Small to Make a Difference* (2019) and her compelling informational videos have reached and mobilized countless numbers. As she says in the video titled "Our Relationship to Nature is Broken": "Those with the most power have the most responsibility, and most of us can do something. What will you do?"

---

[75] https://www.independent.co.uk/climate-change/sustainable-living/greta-thunberg-burger-king-earth-indian-donald-trump-b1850074.html "6 Young Climate Activists making waves, who aren't called Greta Thunberg," Luke Rix-Standing, *Independent*, May 27, 2021.

[76] https://www.newyorker.com/magazine/2021/05/31/are-we-entering-a-new-political-era "Are we entering a new political era?", Andrew Marantz, *The New Yorker Magazine*, May 24, 2021; published in the print edition of the May 31, 2021, issue, with the headline "The Left Turn."

Finally, she stands in solidarity with the climate scientists, never failing to credit those who sounded the alarm first, more than 30 years ago, and those who have worked tirelessly in the meantime, proving and predicting, researching and tracking global warming's many manifestations. She relies on them, too, saying that she writes her own speeches, but "I have a few scientists that I frequently ask for help on how to express certain complicated matters. I want everything to be absolutely correct so that I don't spread incorrect facts, or things that can be misunderstood." When speaking in Brussels in 2019 about the nature of her and her followers' protest, she said, "Unite behind the science. That is our demand."

As she says in her recent video, no comprehensive change will occur unless we "connect the dots." Thinking of Greta Thunberg — and those she stands with, and those who stand with her — makes this necessary task and mobilization seem possible, with a better future just ahead.

James Hansen: Climate Scientist Plus

James Hansen is the leading scientist of the climate crisis. A long-time NASA researcher, Hansen's home field is radiation science. Since climate change is at heart a perturbation of Earth's energy balance, the science of radiation is key to understanding and projecting the effects of increasing greenhouse gases (GHGs). He started calculating the radiative balance of other planets, but then turned his attention to our own home planet.

At the NASA Goddard Institute for Space Studies (GISS), he was the lead developer of one of the first global climate models (GCMs) — the main tool that scientists use to calculate the effects of increasing GHGs on the climate system. In the 1970s, Hansen gathered a small team of modelers at GISS — David Rind, Andrew Lacis, and Gary Russell, among them. Together, they created the GISS GCM.[77]

They began to run simulation experiments to see what the effect of increasing GHGs by humans would have on the climate system. The results were startling. All the simulations showed warming effects that

---

[77] https://data.giss.nasa.gov/gistemp/, GISS Surface Temperature Analysis, National Aeronautics and Space Administration (NASA), Goddard Institute for Space Studies (GISS), 2021.

were amplified by positive feedback effects such as the melting of snow and ice in the high latitudes lowering the reflectivity (also known as albedo) of that region. This change from the white reflective snow to the darker less reflective earth enables more solar radiation to be absorbed at the surface, leading to yet more warming.

The GISS GCM modelers published their first major paper in the journal *Science* in 1981. The paper predicted that if fossil fuel burning continued there would be significant global temperature increase and projected that the warming would be accompanied by increased likelihoods of sea level rise, loss of Arctic Sea ice, and intensified droughts.[78] Those predictions and projections are now coming to pass.

In 1988, Hansen was asked to testify to the Senate Committee on Energy. It was one of the landmark events in the long struggle to have climate change accepted as a "real" global threat requiring urgent action. In his testimony, Hansen stated unequivocally that it was 99% certain that recent warming trends were caused by humans releasing GHGs into the atmosphere from the burning of fossil fuels. At that time, 1988 was the hottest year on record. Hansen shared data with the senators showing that the planet had been warmer in the first 5 months of 1988 than in any comparable period since measurements began 130 years ago.[79]

To explain how increasing GHGs in the atmosphere work to increase the chances of warmer temperatures in the summer, Hansen devised what he called the "climate dice."[80] An ordinary set of dice has two 6-sided cubes with each side having an equal chance of coming out on top. Hansen created a set of cubes basing them on summer temperatures between 1951 to 1980. The dice were "loaded" to reflect the heat-trapping

---

[78] https://science.sciencemag.org/content/213/4511/957, Hansen, J., Johnson, D., Lacis, A. *et al.*, Climate Impact of Increasing Atmospheric Carbon Dioxide, *Science*, 213 (4511), 957–966 (1981); DOI: 10.1126/science.213.4511.957.

[79] https://www.nytimes.com/1988/06/24/us/global-warming-has-begun-expert-tells-senate. html, "Global Warming Has Begun, Expert Tells Senate," Philip Shabecoff, *New York Times*, June 24, 1988.

[80] https://news.climate.columbia.edu/2012/08/06/the-new-climate-dice-the-odds-have-shifted-to-hot/ "The New Climate Dice: The Odds Have Shifted to Hot," David Funkhouser, *State of the Planet*, Columbia Climate School.

effects of the human-caused increases in atmospheric greenhouse gases. Hansen's climate dice had four red sides for hot summers, one blue side for cool summers, and one side for in between. So, when the climate dice are rolled, more summers come out being hot.

The hotter summers related to increasing greenhouse gases that Hansen explained through his climate dice have continued to occur since 1980, with the summer of 2021 breaking temperature records in the U.S. Pacific Northwest. Portland, Oregon recorded 46.6° Celsius (116° Fahrenheit) on Monday, June 28, 2021, setting a new record for the third day in a row.

Working with Sergei Lebedev and Reto Ruedy, Hansen was one of the first to compile temperature records from all over the world and calculate the global surface temperature and its change with time. Thus, he not only modeled the temperature rise mathematically, he was "ground-truthing" it by comparing the model simulations to observations. Temperature is measured at weather stations at the height of about 1.5 meters (5 feet) on land and near the surface of the ocean by instruments in buoys, ships, and satellites. The land and ocean temperatures are averaged together to create the global surface temperature dataset.

The procedure to track global surface temperatures, which Hansen and Lebedev originally set up, is still being carried out today at the NASA Goddard Institute for Space Studies. Since then, other organizations in the U.S. and the UK have developed variations of the procedure and annually publish updates of their global surface temperature datasets as well.

Through the years, Hansen continued his work as a scientist, but he increasingly became an activist, calling out the lack of governmental response to climate change.[81] He was motivated by the slow pace of action and the obfuscation by climate denialists of the science.

Hansen's 2008 paper inspired the name for the activist organization, 350.org. The name refers to the 350 parts per million (ppm) CO_2 that Hansen and his co-authors established as the safe upper limit for carbon

---

[81] https://www.theguardian.com/environment/2018/jun/19/james-hansen-nasa-scientist-climate-change-warning, "Ex-Nasa scientist: 30 years on, world is failing 'miserably' to address climate change," Oliver Milman, *Guardian*, June 19, 2018.

in the atmosphere.[82] Levels were above 418 ppm as of March 2022,[83] indicating that the world is dangerously over its limit and at great risk of major disruptions to come.

Hansen has written letters to numerous world leaders about the dangers of climate change, including to the former UK Prime Minister Boris Johnson. The letter to the UK Prime Minister was prompted by Johnson's decision to open a new coal mine in Cumbria. In the letter to Johnson, Hansen called on him to stop all support for fossil fuels and "earn a special place in history" for tackling the climate crisis. He wrote:

> In leading the UK, as host to the COP, you have a chance to change the course of our climate trajectory — or you can stick with business-almost-as-usual and be vilified in the streets of Glasgow, London and around the world.[84]

Hansen has also participated in many climate protests. One led him to be arrested in 2010 outside the White House during a protest against the Keystone oil pipeline. That long-contested project was finally cancelled by its Canadian sponsor in June 2021.

Retired from NASA,[85] Hansen continues to publish scientific papers but also writes popular books about climate change aimed at a wider audience. These include *Storms of My Grandchildren: The Truth About the Coming Climate Catastrophe and Our Last Chance to Save Humanity*, published by Bloomsbury USA in 2009.

---

[82] https://openatmosphericsciencejournal.com/VOLUME/2/PAGE/217/, Hansen, J. *et al.*, Target Atmospheric CO: Where Should Humanity Aim?, *Open Atmospheric Science Journal* 2, 217–231, (2008), DOI: 10.2174/1874282300802010217.

[83] https://www.co2.earth/daily-co2, Daily CO_2.

[84] https://www.theguardian.com/science/2021/feb/04/top-climate-scientist-warns-pm-over-contemptuous-cumbria-coalmine-plan, "Top climate scientist warns PM over 'contemptuous' Cumbria coalmine plan," Fiona Harvey, *Guardian*, February 4, 2021.

[85] https://350.org/breaking-news-about-old-friend/, Breaking News About an Old Friend, Dr James Hansen's retirement.

In an interview with YaleEnvironment360 titled "For James Hansen, the Science Demands Activism on Climate,"[86] he said:

> I do think that more scientists need to be as clear as they can in communicating the implications of the science to the public...We have a society in which most people have become unable to understand or appreciate science. ... [T]hat's a communication problem which we need to ... alleviate.

When asked by the reporter about being an alarmist, Hansen replied:

> The scientific community is getting more and more concerned about the fact that we're pushing the system beyond tipping points and things are happening. I don't think that I have been alarmist — maybe alarming, but I don't think I'm an alarmist.

In summary, Hansen has been prescient in knowing there was a colossal problem with humans causing changes to the radiation balance of the Earth and understanding how to go about using the best science to investigate it. He did not stop there, however, and went forward boldly to advocate for solutions to the predicament of climate change as well. The leadership of James Hansen indeed sounds the alarm for climate change science, communication, and action across the globe.

---

[86] https://e360.yale.edu/features/james_hansen_science_demands_action, "For James Hansen, the Science Demands Activism on Climate," Katherine Bagley, *Yale Environment 360*, April 12, 2016.

Dimension 5

Finance

Wall Street (New York City, 2008)

Finance

Dimension 5: Finance explores the role of corporations in creating, perpetuating and solving the climate crisis. Like governments with their policies and regulations, businesses with their actions can either accelerate global warming or can help to stop it. Companies from the extractive, industrial, and service sectors all need to set goals for greenhouse gas reductions in their operations plans. Some of these actions can improve their bottom lines because increasing energy efficiency can save money.

Over 400 companies in the U.S. now have goals to achieve net zero greenhouse gas emissions. Far more, however, continue to pursue short-term profits, disregarding with scant consideration the fast-approaching calamity they are hastening. On the positive side, two of the largest international funding agencies, the World Bank and the International Monetary Fund (IMF), have stopped funding coal projects, and major insurance firms have stopped writing policies for projects in environmentally sensitive areas such as the Arctic.

In 2007, Exxon had the largest market capitalization of any company listed on the New York Stock Exchange. In August 2020, it was delisted from the Dow Jones Industrial Average after being a fixture in the index for 92 years, in part because of the trend towards renewable energy. However, the top 60 banks and investment firms have provided $1.8 trillion to the fossil fuel industry in the little more than five years since the Paris Agreement was signed.

The fossil fuel industry is especially culpable. It pays lip service to the growing popular clamor for a viable future, issuing pronouncements setting distant goals to reduce emissions somewhat. Their statements amount to little more than greenwashing, a deceptive marketing technique that spins corporate messages to sound more environmentally friendly than is warranted by the truth. Meanwhile, the industry continues on its way, planning to increase their production capacity by 50% over the next five years.

Coal, oil, and gas companies, some publicly owned and others owned by nation states, are kept in business by generous government subsidies. It is not as if they need the handouts. Exxon's market capitalization is a whopping $265 billion. Aramco, the Saudi Arabian state oil company, launched an initial public offering (IPO) in December 2019 of 1.5% of its stock, which valued the company at $1.7 trillion.

In regard to national policies that can speed GHG reductions, however, only a few countries have implemented carbon pricing, a mechanism aimed at lowering emissions and making polluters share in the monetary burden imposed by climate change. Few, if any, nations have ended the often massive subsidies accorded to their fossil fuel industries.

As of December 2019, over $12 trillion dollars had been divested from the fossil fuel industry by activist organizations and individual shareholders who no longer wish to own stocks in these companies. This divestment has been without any apparent effect. In just the next five years, the industry will invest $1.4 trillion in new production, which will generate emissions that will push the climate's temperature rise well beyond the aspirational target of 1.5° Celsius (2.7° Fahrenheit) and approach or exceed the 2° Celsius (3.6° Frahrenheit) limit set in the Paris Agreement.

Signs of change have been emerging in the last few years. For example, some citizen groups are turning a spotlight on Big Oil. Banks, sovereign funds, pension funds, and international lenders like the World Bank and the IMF, as well as large insurance companies, are also beginning to put checks on the fossil fuel industry through loan constraints and setting of performance standards.

The fossil fuel industry is like an enormous supertanker — slow and cumbersome. However, the pressure to change course is inexorable. Governments often have strong ties to the fossil fuel industry and are loath to commit to change, but the finance sector is more attuned to the intensifying climate change risks to their monetary rewards.

For example, some financial firms have announced that they will begin to require the fossil fuel firms they finance to assess their own exposure to the climate risk inherent in exploiting fuels that produce greenhouse gas emissions. The requirements also include targets for reducing those emissions. Goldman Sachs has taken the unprecedented step of setting its own policies regarding what types of operations it will no longer finance. It has made continued financing contingent on the targets being met by the fossil fuel companies.

Financing for new coal plants is now almost impossible to get in the U.S. and is starting to dry up for small independent fracking operations. At least one major insurance company, Chubb Limited, has stopped writing insurance policies for coal projects. Without insurance, no fossil fuel project can go forward. Because of these types of responsible actions by some in the financial world, there is hope for the Earth and all its living creatures.

NYC Hotels Take up the Carbon Challenge

The Associated Press reported in 2015 that over a dozen famous hotels in New York City had pledged to cut their greenhouse gas emissions by 30% by the year 2025. Those hotels include the Waldorf-Astoria New York, the Lotte New York Palace, the Pierre, the Crowne Plaza Times Square, the Grand Hyatt New York, the Westin New York at Times Square, and the Peninsula New York. By taking energy-saving steps over the next nine years, estimated greenhouse gas emission reductions could be as much as 32,000 metric tons, and cost reductions could be about $25 million.

The hotels have signed up with the New York City Carbon Challenge program and will join other area hotels that have made energy efficiency upgrades, such as replacing electric bulbs with energy efficient LEDs (light-emitting diodes).[1]

Hotels joining the program are required to submit an annual carbon emissions inventory via a web-based NYC Carbon Challenge Reporting Tool platform and share that information through a U.S. EPA Portfolio Manager tool. Challenge participants are also required to review data quality annually using a simple data quality guide provided by the Mayor's office.

Other New York City buildings have already started taking part in the Carbon Challenge program. Police stations and firehouses are now using LED lighting upgrades. There have been heating and lighting upgrades at landmark museums, such as the American Museum of Natural History, the Metropolitan Museum of Art, and the Brooklyn Museum, whose energy costs are underwritten by the City. Battery storage technologies have been installed at Queens Hospital and Jacobi Hospital.

---

[1] https://www1.nyc.gov/, the official website of the City of New York.

Update: In the years since this blog was written, more than 100 entities in the private, institutional, and non-profit sector have taken the Carbon Challenge pledge to reduce greenhouse gas emissions by 30% or more within 10 years.[2] Among those who have committed to achieving that goal are many of the city's largest universities, hospitals, hotels, corporate offices, retail stores, and property management firms. By 2021, 21 had met the goals, and 19 had committed to 50% or greater reductions by 2025. In all, participants had lowered their annual GHG emissions by 580,000 metric tons of carbon, a total that is expected to approach 1.5 million metric tons by the end of the program in 2025.

Divesting from Fossil Fuels

The first major event on climate change for investors and business was the 2016 Investor Summit on Climate Risk held at the United Nations (UN) in NYC.[3] The summit hosted some 500 global investors who, in total, represented an estimated $22 trillion in assets and set forth a new direction in divesting in fossil fuels.[4]

Since that meeting, there has been a surge of information and news reports about investment strategies to divest in fossil fuels because of their direct, negative impact on the environment through climate change.[5] The financial community has seen the value of fossil fuels fluctuate, offering unstable returns. As well, companies have come to realize that investing in sustainable, low carbon assets are seeing stronger returns. Others are stressing the importance of investing in clean energy instead of pouring

---

[2] https://www1. nyc.gov/site/sustainability/our-programs/carbon-challenge.page

[3] https://unfccc.int/news/2016-investor-summit-on-climate-risk "2016 Investor Summit on Climate Risk," United Nations Framework Convention on Climate Change (UNFCC), January 26, 2016.

[4] https://www.ceres.org/news-center/press-releases/500-global-investors-managing-trillions-mobilize-action-wake-paris "500 Global Investors Managing Trillions Mobilize Action in Wake of Paris Climate Agreement," Ceres, January 27, 2016.

[5] https://www.theguardian.com/sustainable-business/2016/feb/13/renewable-energy-investment-fossil-fuel-divestment-investor-summit-climate-change "Have we reached the tipping point for investing in renewable energy?" Bruce Watson, *Guardian*, February 13, 2016.

dollars into fossil fuels. However, there are challenges with any new investment strategy. The American Council on Renewable Energy (ACORE)[6] has illustrated that clean energy investments pose many challenges because the rates of return from different types of renewables have also fluctuated through time.

High-profile companies getting involved in clean energy investments include Walmart, Ikea, Apple, and Google. Groups that have long called for Wall Street to invest seriously in renewable forms of energy include Greenpeace, World Resources Institute, World Wildlife Fund, Rocky Mountain Institute, We Mean Business, and the RE100.

In order for corporations to make meaningful contributions to greenhouse gas reductions, their actions must be documented. The Carbon Disclosure Project (CDP) is a not-for-profit charity that provides independent verification for the climate change actions of investors, companies, cities, states, and regions.[7]

What is a Carbon Bubble?

American sustainability consultant Alex Steffen writes, "When there's a large difference between how markets think assets should be valued and what they are (or will) actually be worth, we call it a 'bubble.'" While scientists are convinced that the Earth's climate is changing and that this change is largely due to human activities, many investors are still unaware of this circumstance. They still place high valuations on the assets of coal, oil, and natural gas companies. This is known as the "Carbon Bubble."

In an article published on Medium.com[8] entitled "Trump, Putin and the Pipelines to Nowhere,"[9] Steffen explains why such valuations are unwarranted. As countries curtail their use of fossil fuels in favor of

---

[6] https://acore.org/, American Council on Renewable Energy (ACORE).

[7] https://www.cdp.net/en, Carbon Disclosure Project (CDP).

[8] https://medium.com/, Medium.com.

[9] https://thenearlynow.com/trump-putin-and-the-pipelines-to-nowhere-742d745ce8fd#. dtqvjfsaf, "Trump, Putin and the Pipelines to Nowhere," Alex Steffen, *The Nearly Now*, December 16, 2016.

renewable sources of energy such as solar, wind, hydro, and geothermal, fossil fuels will diminish in value. Untapped fossil fuel reserves increasingly become, not assets, but liabilities. As Steffens says, if we cannot burn oil, it is not worth much. Other assets besides fossil fuel are poised to be devalued, such as coastal real estate that cannot be insured because of rising sea levels and companies newly vulnerable to climate litigation.

Steffen cites Mark Carney, the Governor of the Bank of England[10] and chair of the Financial Stability Board[11] — the global institution charged with preventing market panics and crashes — who gave a bombshell talk at Lloyd's in 2016. He warned that the carbon bubble would continue to grow and could lead to a global market crisis as big as or worse than the 2007 subprime mortgage crisis. When the carbon bubble bursts, he forecast, trillions of dollars of imaginary assets will quickly disappear.

It comes down to how investors assess risk. If they see their assets headed for a risky future, skittish investors will divest and seek safer investments. The herd will follow, causing share prices to plummet. Fossil fuel companies face a growing threat from loss of investor confidence. They know this and have spent millions of dollars lobbying to deny climate change and humanity's responsibility for it. They have spun the notions that fossil fuels are vital to Earth's current prosperity and that the future growth and well-being they will continue to provide are assured.

In recent years, fossil fuel companies have begun to shift their messages to the media from denying climate change, attacking global climate agreements, criticizing carbon alternatives, and opposing any tax on carbon, to now acknowledging that a mix of energy sources is essential.

While the environmental indictment of fossil fuels is strong, ultimately, the economics will prevail. Eventually, the carbon bubble will burst as the profits of the fossil fuel industry come to be seen as increasingly uncertain. Divestment and bankruptcies will ensue. Several coal companies have already gone bankrupt in the last few years. Industrial-scale solar and wind energy are now cheaper than either oil or gas.

---

[10] https://www.bankofengland.co.uk/, Bank of England.
[11] https://www.fsb.org/, Financial Stability Board.

Experts from the Baker Institute for Public Policy, a non-partisan think tank at Rice University, have warned that the carbon bubble "is one of the biggest threats to the global economy."[12] To manage the deflation of the carbon bubble will entail cutting fossil fuel use and speeding adoption of measures to slow the climate change trajectory. Such measures would, we can only hope, let the carbon bubble deflate slowly enough to allow markets to adjust and avert a global financial crisis.

Playing Both Sides of the Fossil Fuel Fence

In 2017, a report about how the Group of 20 (G20, the main economic council of wealthy nations) governments are pouring more funds into fossil fuels appeared on the Oil Change International site. This group is a research, communication, and advocacy organization that exposes the true costs of fossil fuels and seeks to facilitate future transitions towards clean energy.[13] The group partnered with Friends of the Earth US, the World Wildlife Fund, and the Sierra Club to create the report.

Science has overwhelmingly shown that preventing global warming means keeping temperatures well below 2° Celsius (3.6° Fahrenheit) to ultimately limit warming to 1.5° Celsius (2.7° Fahrenheit) by curtailing the use of fossil fuels. This goal was a major part of the Paris Agreement[14] signed by 195 countries in April 2016 at the United Nations Framework Convention on Climate Change (UNFCCC) 21st Conference of the Parties (COP21).

Alarmingly, many governments who signed the agreement continue to fund fossil fuel projects. The report, entitled "Talk is Cheap: How G20

---

[12] https://www.bakerinstitute.org/media/files/research_document/6b58fc69/WorkingPaper-ClimateRisk-072116.pdf "Climate Risk and the Fossil Fuel Industry: Two Feet High and Rising," Jim Krane, Working Paper, The Baker Institute for Public Policy, Rice University, 2016.

[13] http://priceofoil.org/, Oil Change International.

[14] https://unfccc.int/process-and-meetings/the-paris-agreement/the-paris-agreement, The Paris Agreement.

Governments are Financing Climate Disaster,"[15] shows how many of the G20 nations still fund the fossil fuel industry. They are among the world's largest economies (and polluters), who, in Paris in 2015, had committed to curtail support for fossil fuels in favor of low-carbon alternatives. The report shows that the shift from fossil fuel funding to low-carbon alternatives has yet to happen.

Here are some alarming figures:

According to the "Talk is Cheap" report, the U.S. financed fossil fuels — oil, gas, and coal — to the tune of $6 billion annually from 2013 to 2015, versus $1.3 billion for renewables.[16] Japan is the largest public funder of fossil fuels — $16.5 billion compared to $2.7 billion for renewables. China spent $13.5 billion on funding fossil fuels and only $85 million on funding renewable energy sources. South Korea was third largest funder at about $9 billion compared to only $92 million invested annually in clean energy, with U.S. being the fourth. All together, the G20 nations and the multilateral development banks poured $71.8 billion into fossil fuel production; only $18.7 billion annually — or 15% — supported clean energy (including renewable sources such as wind, solar, geothermal, and hydro). As one of the authors of the report said, "We must stop funding fossils and shift these subsidies."

Update: There is progress. As of 2021, renewable energy sources are about 70% of the total USD 530 billion spent on all new generation capacity. According to the International Energy Agency (IEA), thanks to technology improvements and costs reductions, a dollar spent on wind and solar photovoltaic (PV) deployment today results in four times more electricity than a dollar spent on the same technologies ten years ago.[17]

---

[15] http://priceofoil.org/content/uploads/2017/07/talk_is_cheap_G20_report_July2017.pdf
http://priceofoil.org/content/uploads/2017/07/talk_is_cheap_G20_report_July2017.pdf
"Talk is Cheap: How G20 governments are financing climate disaster," Alex Doukas, Kate DeAngelis, Nicole Ghio, Kelly Trout and Elizabeth Bast, Oil Change International, 2017.
[16] http://www.investorideas.com/news/2017/energy/07051FossilFuel.asp "U.S. Sending $6 Billion to Subsidize Fossil Fuel Projects Abroad," Investorideas.com, July 5, 2017.
[17] https://www.iea.org/reports/world-energy-investment-2021/executive-summary

Can the Pope Work a Miracle with Big Companies?

Some corporate executives met with the Pope[18] in 2018 to talk climate change. In 2015 Pope Francis issued an encyclical on the subject titled "On Care for our Common Home." The world is in the midst of a transition from fossil fuel-based energy to renewable, low-carbon energy sources like wind, solar, and hydro. Prodded by their shareholders, the Oil Majors are beginning to acknowledge that their business cannot continue as usual for very much longer.

However, these are very big companies that have had things their way for a very long time. Even contemplating the need to change is beyond some of them. Those few who do see the need have a great deal of inertia to overcome. They move, if at all, at a snail's pace. However, time is running out for the industry and the Earth's climate. Bold, immediate actions are needed.

Can the Pope spur the powerful corporate leaders into meaningful action? Moral suasion cuts little ice with this crew. Only money talks. It is the only language they understand. There is an argument His Holiness could make in those terms, one that stresses the inexorability of the transition from high-carbon to low-carbon energy. It is under way and cannot be stopped. Big Oil can slow it down, but delay would not avoid the inevitable. Demand for petroleum-based products will peak and then decline. The price of oil will fall below the cost to pump, refine, and deliver it. The vast reserves of untapped oil will become worthless, uneconomic to exploit. The carbon bubble will burst. Coal mines will close, power plants shut down, and whole countries go bankrupt like Argentina.

Saudi Arabia, led by Crown Prince Mohammed Bin Salman, has started to diversify and divest, so may escape the same fate. Norway too

---

[18] http://www.vatican.va/content/francesco/en/speeches/2018/june/documents/papa-francesco_20180609_imprenditori-energia.html "Address of His Holiness Pope Francis to Participants at The Meeting for Executives of The Main Companies in The Oil and Natural Gas Sectors, And Other Energy Related Businesses," The Vatican, June 9, 2018.

is beginning to lower its dependence on oil. Other oil-rich states may not be so lucky. Iran, Russia, Japan, and the U.S. all face major disruption as the world weans itself off oil. The Oil Majors could become the Renewable Majors if they were smart and nimble, but judging by past performance, they are not.

Norway's Sovereign Wealth Fund Saga

Norway's sovereign wealth fund announced its intention in 2018 to divest its oil and gas holdings. A government commission urged them not to, arguing that doing so would lower revenues, not risk. This dynamic between desire to divest in fossil fuels and reliance on North Sea oil to fund its welfare state is a modern-day Nordic saga. In 2014, the Norway Sovereign Fund had already divested from 53 coal companies around the world, including companies in the U.S., India, and China.[19] As a result, the total value of the fund's coal holdings fell by 5% to $9.7 billion. In 2014, the fund also sold its stakes in 59 out of the 90 oil and gas companies in which it holds shares.

In August 2017, a group of academics, analysts, and activists met in the Lofoten Archipelago in Norway and created the Lofoten Declaration.[20] Activists have successfully blocked offshore oil drilling in this site of great natural beauty for decades. The Declaration is an international manifesto calling for the end of exploration and development of fossil fuels for the purpose of mitigating climate change. It calls for divestment in fossil fuels and the phasing out of their use, and a just transition to a low-carbon economy.

By 2021, the Lofoten Declaration has been signed by about 600 activist organizations based in over 70 countries. A key tenet of the Declaration is that early action is required from those countries that have benefited from fossil fuel development:

---

[19] https://en.wikipedia.org/wiki/Government_Pension_Fund_of_Norway, "The government pension fund of Norway", Wikipedia.
[20] http://www.lofotendeclaration.org/, The Lofoten Declaration.

The Lofoten Declaration affirms that it is the urgent responsibility and moral obligation of wealthy fossil fuel producers to lead in putting an end to fossil fuel development and to manage the decline of existing production.

In March 2019, the Norwegian Ministry of Finance recommended divestiture from its oil and gas exploration and production holdings.[21] While it is believed that the Lofoten Declaration has helped to influence the government of Norway to divest from fossil fuel exploration and production, the fund managers said that protecting against risk of declines in the price of oil was the motivating factor. The fund is also retaining investments in energy companies that they believe will play major roles in developing green energy, such as Shell.

As of 2019, guidelines prohibited the fund from investing in companies that produce over 20 million tons of coal annually. The fund announced plans to phase out its investments in over $10 billion in stocks from companies using too many fossil fuels. In hopes of improving the Norwegian economy, the fund is investing in companies that promote renewable energy. The saga continues, as Norway's dedication to the environment conflicts with its reliance on North Sea oil.

Exxon Held to Account for Climate Denial

In an article by Alexander C. Kaufman entitled "Exxon's Climate Denial Set to Face First Public Scrutiny As Legal Woes Mount" published on *Huffpost*,[22] we learn that Exxon is finally getting some well-deserved heat for its considerable part in creating the gathering climate change crisis we find ourselves experiencing today.

---

[21] https://www.forbes.com/sites/davidnikel/2019/06/12/norway-wealth-fund-to-dump-fossil-fuel-stocks-worth-billions-in-environmental-move/?sh=73cfb98748a3, "Norway wealth fund to dump fossil fuel stocks worth billions in environmental move," David Nikel, *Forbes*, June 12, 2019.

[22] https://www.huffpost.com/entry/exxon-mobil-climate-change-denial_n_5c901482e4b0d50544fee0f2 "Exxon's Climate Denial Set to Face First Public Scrutiny As Legal Woes Mount," Alexander C. Kaufman, *Huffpost*, March 19, 2019.

Exxon and the other oil majors have long denied any culpability for causing the consequences of the energy system based on the combustion of fossil fuels.[23] It is altogether fitting that the fossil fuel industry finally be called to account for their long years of promotion of personal automobiles. The industry has disparaged the science detailing the negatives of "car culture" that they knew was correct but that threatened their business model and their hefty profits.

The World's Growth Model is Unsustainable

Steven Chu, former Secretary of Energy, 1997 winner of the Nobel Prize for Physics, and former president of the American Association for the Advancement of Science (AAAS), critiques all world economies as being pyramid or Ponzi schemes, dependent on a growing population and inherently unsustainable. As reported in a *Forbes* green tech article[24] entitled "The World Economy is a Pyramid Scheme" written by Jeff McMahon,[25] Chu proposes an alternative world economic model.

Chu's new model is based on universal education for women and on wealth creation for all, the two established ways to reduce fertility, and thus population growth. Having a new economic model is important because over-consumption of resources and emissions of massive amounts of CO_2 and other greenhouse gases have caused changes to the climate that are increasingly eroding the livability of many of Earth's inhabitants.

There will certainly be issues raised by the transition from a growth model to a sustainable steady-state one. Who farms the land to produce

---

[23] https://www.betterworldsolutions.eu/exxonmobil-gave-millions-to-climate-denying-lawmakers/ "ExxonMobil gave millions to climate-denying lawmakers," Better World Solutions, July 21, 2015; and https://www.theguardian.com/environment/2015/jul/15/exxon-mobil-gave-millions-climate-denying-lawmakers?CMP=EMCENVEML1631 "ExxonMobil gave millions to climate-denying lawmakers despite pledge," Suzanne Goldenberg, *The Guardian*, July 15, 2015.

[24] https://www.forbes.com/search/?q=Green%20Tech&sh=72927b0c279f, Forbes Green tech.

[25] https://www.forbes.com/sites/jeffmcmahon/2019/04/05/the-world-economy-is-a-pyramid-scheme-steven-chu-says/?sh=2aee3e7f4f17 "The World Economy Is A Pyramid Scheme, Steven Chu Says," Jeff McMahon, *Forbes*, April 5, 2019.

the food? Who works in the factories, offices, and shops, and who provides services? If all the young, few as they are, are not needed to work, how do they survive? Who serves in the military defense forces?

A new model is needed because caring for aging populations is starting to bite hard at the economies of Japan, China, Europe, and the U.S. Who cares for the elderly? What happens to the smallest economic unit, the family? An aging population with below-replacement fertility implies an ever-decreasing population, which is itself unsustainable. While some benefit derives from a lower human burden on the Earth's resources, how will societies manage the decline? If a new equilibrium is not achieved, human extinction could occur.

We should listen to Steven Chu, heed his warning that the growth model dominant throughout human history is unsustainable because it leads to irrevocable negative climate impacts. We must begin the gargantuan task of transitioning to a new sustainable and climate-controlled world economic model,[26] a transition for which there is really no human precedent.

Investor Summits: Advancing the Clean Trillion

Key players in the financial world began to gather in 2016 to learn how to invest trillions of dollars in the fight against global warming. The "2016 Investor Summit on Climate Risk: Advancing the Clean Trillion" was billed as the first major event on climate change for investors and businesses since the international climate talks in Paris. The 2016 Investor Summit on Climate Risk was held at the United Nations (UN) in New York City in January, 2016.

One co-host of the summit was Ceres,[27] a large non-profit organization based in Boston, Massachusetts, with an investor network that includes over 180 institutions that manage over $30 trillion in assets. It mobilizes institutional investors and corporations to tackle climate change. It works to address climate change at all levels of business and investment planning and operations. A significant focus for Ceres is advo-

---

[26] https://www.forbes.com/sustainability/?sh=36200f12363b, Forbes Sustainability.

[27] https://www.ceres.org/our-work/climate-change, Climate Crisis, Ceres.

cacy for robust climate disclosure in financial filings and sustainability metrics. Other summit co-hosts were the United Nations Foundation that informs the business community on how to fight climate change and the United Nations Office for Partnerships.[28] The first one-day event hosted hundreds of attendees from around the world to discuss how to put the goals of the COP21's Paris Agreement into action.

Important topics discussed at the 2016 Summit included types of capital available that help meet global climate goals as well as how to invest in clean energy in developed countries. A representative from Ceres said that "in order to limit global warming to 2° Celsius (3.6° Fahrenheit) and avoid the worst effects of climate change, the world needs to invest an additional $44 trillion in clean energy — more than $1 trillion per year for the next 36 years."

Keynote speakers included Ban Ki-moon, former UN Secretary-General; Christiana Figueres, then Executive Secretary of the UN Framework Convention on Climate Change (UNFCC); and Michael Bloomberg, founder of Bloomberg L.P. and Bloomberg Philanthropies, and three-term Mayor of New York City. At this time, Mayor Bloomberg was the United Nations Secretary-General's Special Envoy on Climate Ambition and Solutions.

Also scheduled to address the Investor Summit was Al Gore, former Vice President of the United States, co-founder and chairman of Generation Investment Management, and chairman of the Climate Reality Project,[29] a non-profit organization he founded that focuses on solutions for climate change. Christiana Figueres, then Executive Secretary of the UNFCCC, spoke, along with French Environment Minister Ségolène Royal, Mindy Lubber, the President of Ceres, Bank of America CEO Brian Moynihan, and leading investors from ABP, AXA, BlackRock, CalSTRS, and the New York State Common Retirement Fund.

---

[28] https://www.un.org/partnerships/, United Nations Office for Partnerships.
[29] https://www.climaterealityproject.org/, Climate Reality Project.

At the time of the first Investors Summit, some initial positive steps had been taken in the investment world. New York State Comptroller Thomas DiNapoli had already put $2 billion into the state's public employee pension fund, a fund created by Goldman Sachs made up of companies with smaller carbon footprints. The state's fund also invested about $1.5 billion over its already-invested $500 million in its Green Strategic Investment Program.[30] This program was for companies operating with the emissions goals of COP21. Low-carbon funds that started with seed money from the UN have surfaced under well-known asset managers such as Black Rock, State Street, and Mellon Capital in the U.S., and Amundi Asset Management and Northern Trust in Europe.

The Ceres-sponsored UN Investor Summit on Climate Risk has been followed by a set of similar convocations focused on the private sector's role in climate change action. It convened again in 2018, with more than 450 institutional investors, corporate leaders, state pension fund managers, and major non-profits in attendance. The meeting was deemed a great success. Awareness of the need for and urgency of significant public and private financing was heightened. Attendees heard the estimate of the Organisation for Economic Cooperation and Development (OECD) that $6.3 trillion was needed annually until 2030 to meet the infrastructure needs that climate change would impose.

At the 2018 Investors Summit, several new initiatives were added to existing ones to facilitate, coordinate, and support efforts to promote bigger stakes in renewables and clean energy infrastructure so as to align their investments with the Paris Agreement goals. For example, the new Investor Agenda built on the Climate Action 100+ begun in Paris. It joins 225 financial institutions controlling $26 trillion in assets to exert pressure on major greenhouse gas (GHG) emitters — and themselves as asset owners and managers — to report their climate-related risks. These efforts have borne fruit with investors, increasing interest in the stocks of companies, mutual

---

[30] https://www.osc.state.ny.us/files/about/pdf/green-strategic-investment-program.pdf, Green Strategic Investment Program Fact Sheet.

funds, and exchange-traded funds (ETFs) that score well on performance ratings of Environmental, Social, and Governance (ESG) measures.

This trend exploded in 2020, fueled by the growing awareness of a deteriorating climate and the need for action to slow those changes. A more unpredictable climate brings risks to companies who do not heed present and clearly foreseeable changes. It also brings opportunities to those companies that recognize the risks and nimbly adjust their products, processes, and practices.

Several entities made major commitments at the 2018 Summit, but "major" is a relative term. A $2 billion investment or divestment is not trivial, but it pales when the need is for trillions. The mechanisms fostered by Ceres and its partners at the Investor Summit ponied up billions, but the need is for over $6 trillion per year. So far, the commitments have never come anywhere near that goal. The lack of progress on approaching the Paris targets is stark evidence of the shortfall.

The principal achievement of the UN- and Ceres-sponsored Investor Summits and other such summits is in raising awareness, not funds. Under capitalism, profit is the measure of success, and the time horizon of most business executives is measured quarterly, not annually or in decades, or longer. Sustainability remains a foreign concept to most capitalists. Their short-term focus blinds them to realities of climate change that scientists, academicians, and activists have no difficulty discerning. If you have spent your entire career trying to find more petroleum, drill more wells, produce more barrels of oil, sell more gasoline, and make more profit, it is very difficult for you to understand what Al Gore or Greta Thunberg is upset about.

It is to Ceres' credit that it has worked diligently to chip away at that mentality, with moderate success. They have been aided by an obviously deteriorating climate, by the efforts of activist investors like Al Gore, and by revelations of the multibillion-dollar, multi-decade misinformation and disinformation campaigns funded by the coal and oil industries. These campaigns touted the benefits of the lifestyles enabled by fossil fuels while challenging, denying, and disparaging the validity of science warnings of dire environmental consequences.

Gradually, more people have noticed that their climate is changing adversely, and they are coming to realize that the energy path they are on holds significant risks of it getting worse and worse. As this awareness grows, social groups are actively confronting corporations and criticizing their widespread indifference to the environment and their past and ongoing damage to it. Members of the public are also urging institutions with which they are affiliated to divest from the fossil fuel industry. They have met with increasing success. In 2014, $52 billion had been extracted; as of September 2020, the total was $14 trillion.[31] Individuals are also removing their personal stakes in the stocks of fossil fuel companies.

All in all, a great deal of progress has been made in raising awareness of both the facts of climate change and its causes in the private sector. However, relatively few corporations and the national governments to whom they provide taxes have been willing to expend the capital on climate change that disrupting long-established arrangements would entail. Workers would lose jobs, communities would lose livelihoods, companies would go out of business, and nations would lose tax revenues needed to run the government, maintain the armed forces, and pay the politicians. Most politicians, like most corporate executives, have a short-term focus. Neither can easily see that it is in their long-term interest to address the looming climate catastrophe sooner rather than later.

Momentum is building for private-sector action on climate change. The March 22–25, 2021, Ceres Conference and the Climate Investment Summit in September 2021[32] were important venues for the expansion of corporate commitments. Day Two of President Biden's Climate Ambition Summit on April 23–24, 2021, brought Bill Gates and corporate leaders together to up their climate change action pledges. The time for the private sector to fulfill their leadership role in climate is now.

---

[31] https://www.statista.com/statistics/1090801/value-fossil-fuel-divestments-worldwide/, "Value of fossil fuel divestments worldwide in 2014 and 2020," F. Norrestad, Statista, December 9, 2020.

[32] https://www.climateinvestmentsummit.org/, Climate Investment Summit.

Put Your Investments on a Low-Carbon Diet

Would you invest in funds that do not include fossil fuel companies? The group, Fossil Free Funds,[33] analyzes the climate impacts of thousands of U.S. mutual funds and exchange-traded funds (ETFs) and shares data on the companies' fossil fuel investments and carbon footprints in easy-to-track formats. Individual investors can find out if their money is helping or hurting the climate. This group started in 2012 tracking 10 companies, and in 2021, tracks over 3,000.

To consider investments that are climate friendly, Fossil Free Funds advises the following:

1. Check out mutual funds and exchange-traded funds (ETFs) on the Fossil Free Funds website.
2. Learn about fossil fuel holdings, carbon footprints, and clean energy strategies of funds you either own or are considering investing in.
3. Look for investment options in funds that align with your values.
4. Consult with your financial advisor about investing in climate-friendly assets.

Regarding green investments, it seems sensible to look at the bigger, older firms with longer track records on green funds first, e.g., Dimensional Fund Advisors, Parnassus Investments, and Green Century Funds. The latter has been doing investing with planetary values for 30 years. All seem to have both U.S. and international funds. Each fund company defines its universe a bit differently, as to what it includes or excludes in its sustainable portfolios. These are, of course, personal decisions for each potential investor.

Market capitalization, commonly called "market cap," is the value of a publicly traded company's outstanding shares. Market capitalization is calculated by multiplying the number of outstanding shares by the share stock price of each share. "Large cap" refers to a company with a market capitalization value of more than $10 billion. "Mid cap" companies are

---

[33] https://fossilfreefunds.org/, Fossil Free Funds.

capitalized between $2 billion and $10 billion, and "small cap" companies between $300 million and $2 billion.

An analyst from Morningstar hedges on green funds, saying investing in a large cap low-carbon fund instead of a "regular" large cap fund probably would not have much monetary effect on a well-diversified portfolio. If oil prices surge, you would see some relative losses without fossil fuel stocks; if the transition to renewables starts impacting oil stocks, as it probably will over the mid- to long-term, you would probably gain.

In short, for personal investors, monitoring assets for fossil fuel positions and carbon footprints is a good way to get involved in climate change action.

Conclusion

We Are Waking Up to the Climate Crisis — Now What?

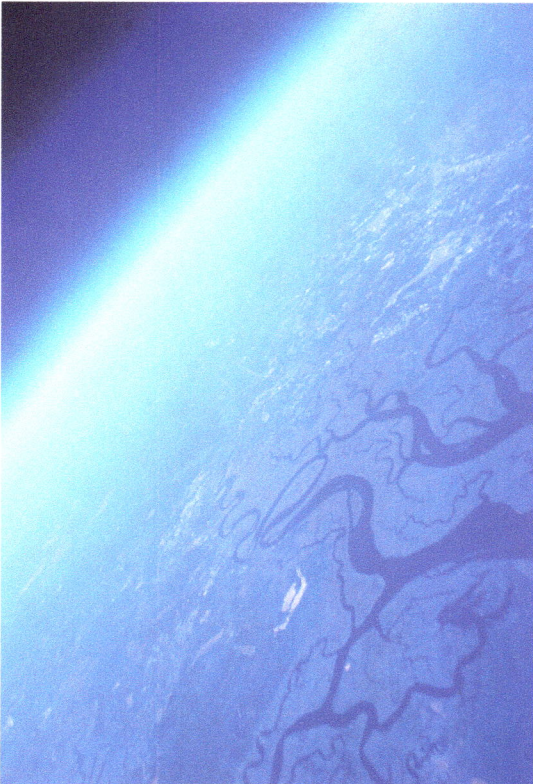

A river system (United States, 2009)

We Are Waking Up to Climate Change — Now What?

In the late 2010s, the climate change issue came to be known as a "crisis." The adoption of this designation was a crucial step in the evolution of public awareness. There is now much greater attention on taking action on all scales to resolve the climate crisis. Work has begun on many fronts, but much remains to be done. (See Appendix 3 for organizations that are engaging in climate change around the world.)

At the global scale, creating a climate-safe world revolves around four main tasks. This is the planetary "to-do list" that is being undertaken internationally by the United Nations Framework Convention on Climate Change. The first task addresses its root cause by reducing overall emissions of greenhouse gases from the burning of fossil fuels. The second and third tasks build up carbon on the Earth by restoring carbonaceous organic matter to depleted soils and reforesting lands to augment carbon stores in trees. The final task recognizes that some climate change will occur no matter what we do to tackle its causes and that the world must adapt.

How do we get started — as individuals and as members of society? There are many ways for us to take action in the fight against climate change in our homes, at our dinner tables, and by our transportation choices. These actions can signal our investment in the issue and help build broader and deeper commitment to climate change solutions in our communities.

Beyond individuals, now is the time for all actors — universities and research labs, corporations, governments, and civil society — to assess their carbon footprints, commit to reducing greenhouse gas emissions, and take action to help solve climate change. Considering the five key dimensions of the crisis — energy; weather and climate; consequences for nature and people; laws and leaders; and finance — the movement forward must be resolute on all fronts.

To-Do List for the Planet

New Year's sun and footprints in the snow (Maine, 2010)

The Four Tasks

Put simply and collectively, the climate crisis presents the world with four global-scale tasks for its planetary to-do list. These are the primary "top-down" actions to be undertaken through international agreements and carried out by national programs.

1.   Curtail greenhouse gas emissions.
To stop the deterioration of our climate and our Earthly habitat, we must stop relying on the energy that comes from burning fossil fuels. That energy comes with emissions that trap heat in the atmosphere, heating the planet and disrupting our complex climate system. Our civilization depends on energy, but for civilization to be sustainable for the long term, we must forego energy from coal, oil, and natural gas, and instead get more and more of our energy from carbonless energy sources, especially wind and solar energy.

2.   Restore organic matter in depleted soils.
Most crops grow in soil, a mix of minerals, organic material, living organisms, gases, and water. Soil contains nutrients that plants need to grow. Growing is an extractive process that can leave the soil depleted if erosion occurs or if good management practices are not followed. Much of the world's cropland is severely depleted, requiring massive applications of fertilizer to ensure adequate crop yield. These applications are generating massive toxic runoffs.

After the crop is harvested, the leaves and stalks ideally are left in the field, and/or the field is protected by a cover crop for a season. This allows the plants to return the nitrogen they drew from the soil while growing, as well as the carbon dioxide that the plants capture from the atmosphere and convert into carbon compounds as they grow, that then decompose. Fertile soil is thus a sink for carbon. Caveats are in order, though, because the stored carbon must remain in the soil continuously and because the soil carbon sink is finite.

Unfortunately, the world's arable land has been eroded, over-farmed, and poorly managed. Regenerative farming is beginning to find widespread adoption in many parts of the globe. The practices are good for the soil, the crops, consumers, the environment, and the farmers' bottom line. In the United States, President Biden is shifting agricultural policy from the Commodity Credit Corporation toward paying farmers to shrink their carbon footprint, and Representative Chellie Pingree of Maine has introduced the Agricultural Resilience Act in the House.

3. Reforest cleared lands to build up carbon stores.

Trees, individually or collectively in forests, perform a vital service for humans. Plants, and especially trees, absorb carbon dioxide and other heat-trapping gases from the atmosphere and, through the process of photosynthesis, convert energy from sunlight into oxygen and plant tissue. Trees thus create the conditions that enable humans to survive and thrive. They create what has been called a "Goldilocks" environment — not too hot, not too cold, just right for humans. Trees have provided humans with the ambient temperature we need and maintained it with only slight fluctuations throughout the entire lifespan of *homo sapiens*.

However, in the roughly 200 years since the beginning of the Industrial Revolution, humankind has upset that fine balance. We have burned billions of tons of the fossilized remains of ancient plants that time has changed into coal, oil, and natural gas. Once burned, the carbon in those once-living plants is released into the air, where it traps some of the heat from the sun, preventing it from escaping back into space, and heating up our climate, nudging us out of our comfort zone.

At the same time as we have mined and burned many billions of tons of fossil fuels, we have cut down millions of square miles of forests to create pasture land for cattle to graze on, or crop land to grow feed for cattle. Only a little is directly used to grow food for humans. Without the forests to remove CO_2 from the air and replace it with oxygen, the imbalance has become more pronounced, and our climate has heated so rapidly that we are facing temperatures not only uncomfortable but life-threatening.

It is essential that we reforest the lands we have denuded of trees for agriculture or cattle raising. Once reforested, we need to make sure that the trees are not cut down again. We also need to realize that the trees eventually reach a carbon equilibrium and thus are a finite sink. And we need to stop cutting forests down in the first place!

4. Develop resilience to climate change now and in the future.

Resilience may seem like a new word because it is now in vogue. But it is actually based on a Latin word and first used in English several centuries ago. Its Latin root means to jump back, and that definition has evolved into the ability to bounce back from adversity. It also implies being prepared to meet whatever comes your way.

These days, with the climate becoming more volatile and less benign, many people are embracing resilience as a way to cope with increasingly frequent, increasingly severe extreme weather events. Rather than being taken by surprise when a storm or flood hits, people are learning to anticipate trouble before it strikes. In the U.S., Floridians do not have to venture out for plywood to board up their windows at the last minute, as they are already stocked. Baltimoreans do not have to strip the store shelves of "white goods" (milk, bread, and toilet paper) before a storm, because they have bought ahead.

More and more people have bought generators as insurance against an electrical outage. More and more people store water in tubs and pails, and keep supplies of drinking water, canned goods, and food that does not need refrigeration.

All are examples of resilience — planning ahead, being prepared, and bouncing back once the worst has passed. There is now a growing movement for bouncing "forward," i.e., transforming human and utilizing natural systems to be even better at responding to climate extremes than before.

Taking Steps

Onward and upward (Maine, 2012)

Ways to Take Action

What can we, in our day-to-day lives, do to help solve the climate crisis? There are many ways to reduce our carbon footprints. Our own and our friends' commitments can help to motivate community action. Turning out the lights when we leave a room lowers our energy usage so the carbon emissions we cause decrease. Recycling paper saves carbon stored in trees, and curtailing use of plastic helps, too, because plastics come from fossil fuels. Scaling down meat consumption can eventually change the economics of the cattle industry, which uses about 20–40% of the Earth's land area for pasturing cattle. This is a main reason why forests are cut down, resulting in the release of large quantities of carbon dioxide. Walking or bicycling rather than driving to do local errands shrinks our carbon footprints. Flying less often can pare them even more.

Here are ways that each of us can take action to solve climate change. Such actions can create a social signal for others to join in.

Transportation

Walk and bike more, drive less.

The exercise will do us good, and we will pollute less and emit fewer greenhouse gases. Start with short walks or bike trips, then gradually extend your range.

Work from home if we can.

By not commuting to work, we will save time, money, and emissions.

Meet virtually more, fly less.

Virtual meetings save the considerable expenses of business travel and the emissions generated getting to the meeting site and returning home.

Combine shopping trips, errands, and deliveries.

We can save a lot of time, gas, and emissions by planning ahead before hopping in the car. Coordinating with a friend or partner can save some entire trips.

Do more remote learning.

Take advantage of remote learning at our own pace. Connect with friends and classmates online too.

Think about ditching gasoline-powered SUVs.

Sports utility vehicles (SUVs) powered by gasoline are popular cars in America and bad for the environment. How often do we go off-road, or use the extra cargo space? We will get a lot better mileage with a sedan, save a lot of money on gas, and if we go for a hybrid or an electric vehicle, we will feel good that we are not polluting the planet nearly as much.

Energy

Switch bulbs to LEDs to save money and electricity.

Lightbulbs have come a long way since Edison invented them in 1879. The incandescent bulbs he developed have given way first to compact fluorescent lamps (CFL) bulbs and now to light-emitting diode (LED) bulbs, which are up to 90% more efficient than incandescent bulbs. They last a lot longer and use much less electricity.

Turn out lights when leaving a room.

Make it a habit. There is no point in lighting an empty room. It will lower our electricity bills, and fewer greenhouse gases will be produced.

Set thermostat a few degrees lower in winter, higher in summer.

In winter, layer up. We will be just as comfortable in a sweater or a sport coat. In summer, dress light. A small fan uses a lot less energy than an air conditioner.

Consider solar panels or a geothermal pump.

Heating and cooling our homes takes a lot of gas, oil, or electricity. It is costly and harms the climate. Solar presents several alternatives. One is to put panels on our roofs that we either lease or buy through monthly payments over several years. Leasing is often a no-money-down option, often with free installation. Several states are pioneering community solar, with a local solar farm supplying electricity to neighborhood homes. Many utilities now offer electricity produced from renewable sources at prices comparable to or even lower than the customary fossil-fuel-sourced electricity. Geothermal pumps are still quite rare but are gaining in popularity due to government incentives.

Upgrade windows to cut down drafts and heat loss.

Windows used to be a single pane of glass. Now they are double or even triple-paned, with a thin layer of air in between that blocks heat from getting in or out. And if windows are old, they probably no longer fit snugly, so they let heat go out the window or let cold drafts in.

Use power strips to turn TVs on and off.

Many TVs have an instant-on feature, which is really a circuit that is always on even when you have turned the set off. A power strip defeats this feature and saves you money and the planet some carbon emissions.

Buy less plastic, especially single-use items.

Plastic is made from and with fossil fuels. Most plastic cannot be recycled. Moreover, plastic takes years to degrade, filling up waste dumps and landfills. Much of it ends up polluting beaches and the oceans. Without thinking, we throw away plastic bags, bottles, straws, utensils, and Styrofoam cups. But they hang around and around. When they finally degrade, they break into microparticles that float in the air, and have been found in both the Arctic and Antarctica. They have even been found in the food we eat and the beer we drink. Scientists have identified several toxic effects of inhaling and ingesting these tiny particles. They are still studying the harm that the particles cause, but the information learned so far has already given us good cause to be worried.

Kitchen and Dinner

Take reusable tote bags when we shop.

Use washable bags instead of flimsy plastic bags that most shops give out, which cannot be recycled or reused and end up cluttering our homes, a landfill, or the ocean. It seems like a small thing, but the average American family of four takes home 1,500 plastic bags a year, and virtually none are recycled. However, cloth totes made of cotton use a lot of water to produce, so there are trade-offs. Totes made of recycled plastic bags and bottles may be the way to go.

Drink more water, less cow's milk.

Dairy cattle are less bad for the climate than beef cattle, but drinking cow's milk still has an environmental cost. Many plant-based alternatives are now marketed — oat milk, coconut milk, almond milk, soy milk, and others. Try them all, and stick with the one you like best. Plant-based milks have some climate cost, but their production emits fewer greenhouse gases than cow's milk. They are a bit more expensive than "real" milk, but worth it to lower our carbon footprints.

Eat more grains, beans, vegetables, and fruits, less meat.

Whole forests have been cut down to become rangeland for beef cattle, losing the carbon-capturing capacity of trees. Other cleared land is planted in feed crops for cattle such as soy, alfalfa, or corn. Cattle convert the grains into the red meat protein we eat very inefficiently. A cow must eat about six pounds of corn to make one pound of meat. That meat has health consequences, too, raising our risks of heart attacks and cancer.

Go all-electric for our kitchens.

All electric appliances with renewable energy sources are the way to go in our kitchens. As we transition to all-electric appliances, microwave ovens can play an important role. They use electricity, which has a carbon footprint unless our energy comes from a renewable source. However, microwaves are better than gas ovens. Gas ovens burn natural gas (mostly methane), polluting the air of your house, and shortening our lives and our families'. So, if you have a gas stove and gas oven, consider swapping them for electric ones. Our kitchens will be healthier and better for the Earth.

Buy locally grown produce.

Farmers markets are becoming more and more ubiquitous. Not only does it feel great to be connected to the hardworking people who raise our food, consuming regional and seasonal food can reduce greenhouse gas emissions, if they are grown efficiently. Greenhouses that use a lot of fossil fuel to raise vegetables may not be good for climate change, even if they are right near where we live. Cold storage also takes a lot of energy. However, when we keep these considerations in mind, fresh, farm-grown food from our farmer neighbors can be a great way to reduce our GHG footprints.

Reduce food waste.

America wastes about 40% of all the food it produces, much of it at the household level. It "costs" carbon emissions to grow the food, harvest it, process it, and transport it to our supermarkets and finally to our homes. Reducing the amount of food that we waste thereby reduces GHG emissions.

Become a smarter shopper.

To reduce the food we waste, plan ahead, sketch out menus for the week, make a list to take to the markets when we shop, stick to it, and resist the temptation to buy more than we need or items that are not on our list.

Start composting.

If we live in a home, not an apartment, we can do our own recycling by composting our food scraps. The compost will enrich our garden's soil and reduce the methane emissions from waste dumps. It is cheap, easy, and satisfying to do. Learn all about it on the web.

Run dishwasher only for full loads.

Heating the water to wash the dishes uses emission-producing electricity, so waiting until we have a full load to run the washer shrinks our carbon footprint and lowers our electric bill.

Turn water-heater temperature down.

Everybody likes a hot shower, but if we have to mix in a lot of cold water to get the temperature just right, we can save money and carbon emissions by turning down the control on our water heaters. If you have a gas water heater, think about switching to an electric one, or better yet, a heat pump that circulates heat from underground. Look into whether heat pumps can be installed where you live.

Clothes and Closet

Buy sturdier clothes, wear them longer, and learn to mend.

The fashion industry has a very large carbon footprint. Its business model is to convince us that we need to buy more of their products to keep up with the latest style or trend. The model is predicated on selling us more clothes than we really need, so we can keep up with the latest trends and look and feel cool. The result is overstuffed closets and lots of discarded outfits passed on to thrift stores or sent to the dump. Clothes manufacturers often produce more clothes than they can sell before "what's hot" has moved on. The result of "fast fashion" is tons of clothes deposited in landfills and waste dumps, which generate vast quantities of methane, one of the biggest contributors to global heating.

Wash fewer, fuller, cooler loads in washing machine.

You would be surprised how much energy, GHG emissions, and money we can save by selecting the warm setting instead of hot, or cold instead of warm, and only running the washer when we really have a lot of clothes to wash or when we are about to run out of clean ones. Our clothes will get just as clean, and they will last longer before fading or fraying.

Dry clothes outside if you can.

The dryer uses a lot of electricity and is hard on our clothes. If we can dry them outside, we will be saving money and the planet. Plus, they will smell fresher, and they would not have lost any fibers that show up as lint in our dryer's lint trap. My mother always dried our family's clothes outdoors in the spring, summer, and fall, in New York State where I grew up.

Activism

Vote for climate-aware local, state, and national candidates.

Get involved in politics at all levels — local, county, congressional district, state, and national. Inform yourself about issues and candidates. Make our views and our voices heard. Listen to the views of others. Think about our feelings, values, and concerns. Form opinions and express them. If someone disagrees, try to understand the other's point of view, consider it, and decide whether to modify our own. Debate and defend our views civilly. Everybody is entitled to their own opinion but not to their own facts.

Call, write, or e-mail your Senators and Representatives.

Our elected representatives are there to serve us. They work for us. Tell them what your concerns are, and what stances you want them to take on issues that concern you. They may not always vote the way you want them to, but remember how many times that has happened when the next election day comes. Be vocal about how much we care about having a climate that lets us live and work comfortably and productively. Communicate your concern for ensuring that your children and theirs have a safe and healthy place to live. Insist that they care as much about the climate as you do.

Share concerns with family, friends, and neighbors.

Talk with everybody you know — and some you do not — about climate change. Everybody's life is full, so they may not be tuned in to the climate. By showing how much you care about how the changing climate affects everyone and everything, you will put climate "on their radar." If we all care, our government representatives will have to act. If they do not, we will get ones who will.

Form or join a local group advocating for climate action.

Once you connect with a few others who care as much about the climate as you do, join the group they are in or form one with them if they are not affiliated with any groups. It is great to hang out and engage with like-minded folks while helping to ensure your concerns have an impact. In unity there is strength. In numbers there is strength. The goal is for everyone to care about sustaining the Earth, our only home, so that it remains a place that welcomes us, nourishes us, and lets us flourish for many generations to come. (See Appendix 3 for organizations that are working on climate change around the world.)

The Path Ahead

Where To From Here?

The climate crisis is a planet-wide existential threat that requires urgent action to ameliorate the worst effects that will happen if we do not act collectively now. Millions will suffer or die from prolonged extreme heat, and frequent fierce storms will bring hurricane force winds and torrential rains causing floods that inundate fields, towns, even cities. (Remember what Hurricane Sandy did to New York, what Hurricane Harvey did to Houston, or what Hurricanes Maria and Fiona did to Puerto Rico?)

We can probably defuse the climate crisis, but we cannot stop it entirely. To slow down the inevitable will require both a national effort akin to the moon mission or the mobilization mounted to fight World War II. Internationally, the United Nations must coordinate similar national efforts from all the signatories to the 2015 Paris Agreement. To date, there has been more talk than action, and many countries are lagging badly in meeting even the insufficient targets they committed to in Paris.

All countries need to realize that the existential threat is global and that the climate crisis affects everyone. Local concerns pale in comparison with the imminent threat of societal collapse in the face of choking pollution, killing heat, widespread hunger due to crop failures, and hordes of climate refugees fleeing their flooding coastal cities, or certain death at home given the breakdown of agriculture, the economy, and social order. The UN Secretary General and the Pope have both deplored the inaction of nations and implored them to act with alacrity. The eminences and the scientists have all spoken. Now it is time for citizens of the world to act to force the powerful of governments and corporations to collaborate to end the threat of impending climate catastrophe.

When an existential crisis is truly global, sovereignty may have to bend. Likewise, capitalism may have to accept some constraints on its profit-making freedom of action. Brazil should not be allowed to burn down the Amazon rainforest. Canada should not be permitted to develop the tar sands of Alberta. Russia should never exploit the huge West Siberian oil reserves. Australia should not be able to give permission to the Adjani mines to dig up and export billions of tons of coal to China. Indonesia, New Guinea, and Nigeria, among others, have oil deposits they are aching to develop, and even Saudi Arabia's Aramco wants to expand its already world-leading oil production capabilities.

The 60 largest banks in the world have lent \$3.8 trillion to the fossil fuel industry for its exploration and production projects since the Paris Agreement. Further investment in such activities must come to a swift halt, no matter how profitable to the financial institutions in question. Popular activism has begun to shame, picket, and boycott them, and has already had some effect. Demonstrations, protests, and pressure at Annual Shareholders Meetings must continue and escalate.

However, it is an immense challenge to wean the world away from dependence on fossil fuels. The war in Ukraine and Europe's attempts to break its reliance on oil and gas from Russia highlight this. But we must hold fast. The climate crisis is real; it is existential; it is global; and it is not only imminent but already upon us.

To analyze future big power politics without giving due consideration to its probable effects is to ignore the 900-pound gorilla in the room. If we do not start curbing the emissions from burning fossil fuels now, by mid-century, just about 30 years from now, global temperatures are projected to rise by 2.0° Celsius to 3.2° Celsius (3.7° to 5.7° Fahrenheit).[1] Sea levels will have risen enough to cause flooding every full moon in many coastal cities around the globe, resulting in migration to higher land, displacement and failure of many businesses, and sharp falls in coastal real estate values.

Extreme weather events will become more frequent and more extreme. Heat waves will be hotter and last longer. The poor, the elderly, the infirm, and infants will be most affected. Some will die. Droughts will be deeper and last longer. Crop yields will fall, food prices will rise, trade in staples will be restricted, many people will be hungry, and some will starve. People will protest these conditions, political responses will be deemed inadequate, and governments will fall.

Before that happens, however, there will be at least one, but probably several attempts to forge new institutional structures to modify, supplement, or replace the ones developed at Bretton Woods after World War II, to

---

[1] High-end SSP5-8.5 scenario; Baseline 1995-2014; IPCC AR6 Assessed Global Warming; IPCC, 2021: Summary for Policymakers. In: Climate Change 2021: The Physical Science Basis. Contribution of Working Group I to the Sixth Assessment Report of the Intergovernmental Panel on Climate Change [Masson-Delmotte, V., P. Zhai, A. Pirani, S.L. Connors, C. Péan, S. Berger, N. Caud, Y. Chen, L. Goldfarb, M.I. Gomis, M. Huang, K. Leitzell, E. Lonnoy, J.B.R. Matthews, T.K. Maycock, T. Waterfield, O. Yelekçi, R. Yu, and B. Zhou (eds.)]. Cambridge University Press, Cambridge, United Kingdom and New York, NY, USA, pp. 3–32, doi:10.1017/9781009157896.001

forestall a complete collapse from occurring. At Bretton Woods, New Hampshire, in July of 1944, 730 delegates from 44 Allied countries met and developed the International Monetary Fund (IMF). The goal was to create a stable global economic system that was based on mutual cooperation. Such coordination is again needed now to create a new climate-safe world order.

Just as kingdoms supplanted the original coalescence of humans in city-states and nation-states succeeded feudal fiefdoms in the early Middle Ages, a shift to a new form of social and political organization is indicated. It is not inevitable, but if humankind fails to achieve at least a nascent form of supra-national, life-oriented, equitable, and sustainable *modus vivendi* by summoning the political will to overcome our climate inertia, unwillingness, and inability to change, then the present world order is likely to fail.

Climate scientists are 97% united in their assessment that conditions will only deteriorate further if humankind as a whole does not cease emitting greenhouse gases. They have determined that to avoid the most calamitous effects of a climate turned hostile those emissions must become net zero by 2050, only 28 years from now.

The 2015 Paris Agreement committed nearly every country to limit their emissions to meet that goal. However, most countries are lagging badly in meeting their commitments. Time is now the enemy, a resource we have wasted since climate scientist James Hansen first warned the U.S. Congress of the danger more than 30 years ago. Now we must act urgently and comprehensively.

And most importantly, we must put aside our differences and unite so that all people work together to save ourselves from an uninhabitable Earth.

A lot has been done. Yet there is much still to do. Let's keep waking up and get on with it. We need to act as if everything that matters is at stake — because it is.

*****

There is an expression in Spanish I learned when I was in the Peace Corps that can be adapted to the climate crisis and what we can do about it:

Pon tu granito de arena en el mundo de tus niños.

In the case of the climate crisis, it means:

Do your part to save the climate that your children will inherit.

Appendices

Appendix 1

ClimateYou.org Sources

The blogs and webpages below are the original sources for the contents of this book. They may be found at www.climateyou.org. Entries for each section are in chronological order of publication date on the website. See footnotes for other sources.

Dimension 1 Energy

Coal, Oil, and Gas

The Questionable Future of Shale Gas, July 10, 2018
https://climateyou.org/2018/07/10/the-questionable-future-of-shale-gas/

Fracking: Biggest Cause of Massive Methane Rise?, August 16, 2019
https://climateyou.org/2019/08/16/fracking-spikes-methane-but-not-for-long-by-climateyou-senior-editor-george-ropes/

Is There an End to Using Coal?, December 18, 2019
https://climateyou.org/2019/12/18/our-take-is-there-and-end-to-using-coal-by-climateyou-senior-editor-george-ropes/

Coal is Dying, and So Are We, January 27, 2021
https://climateyou.org/2021/01/27/coal-is-dying-and-so-are-we/

Energy Sources in Transition

Using More Renewable Natural Gas, November 8, 2015
https://climateyou.org/2015/11/08/using-more-renewable-natural-gas/

Living off the Grid: The New Normal, August 5, 2017
https://climateyou.org/2017/08/05/living-off-the-grid-the-new-normal/

What Does Wind Power Have? Jobs, Jobs, Jobs!, August 22, 2017
https://climateyou.org/2017/08/22/what-does-wind-power-have-jobs-jobs-jobs/

Bury the Power Grid, September 18, 2017
https://climateyou.org/2017/09/18/our-take-bury-the-power-grid/

Shell Maps Radical Plan for Its Energy Future and Ours: HYDROGEN, March 29, 2018
https://climateyou.org/2018/03/29/shell-envisions-a-carbon-free-future-but-how-do-we-get-there-by-climateyou-senior-editor-george-ropes/

Stanford Researchers Say Net-Zero Emissions by 2050 Is Feasible, March 18, 2020
https://climateyou.org/2020/03/18/our-take-stanford-researchers-say-net-zero-emissions-by-2050-is-feasible-by-climateyou-senior-editor-george-ropes/

Denmark Surges in Renewable Electricity Generation, March 18, 2020
https://climateyou.org/2020/03/18/denmark-surges-in-renewable-electricity-generation-by-climateyou-senior-editor-george-ropes/

Nuclear Energy — Where Are We Today?, June 2, 2021
https://climateyou.org/2021/06/02/nuclear-energy-where-are-we-today/

Transportation

Hydrogen Fuel Cell Vehicles Become a Reality, January 20, 2016
https://climateyou.org/2016/01/20/hydrogen-fuel-cell-vehicles-become-a-reality/

Free Two-Day Shipping and Climate Change, November 20, 2017
https://climateyou.org/2017/11/20/our-take-free-2-day-shipping-climate-change-by-climateyou-senior-editor-george-ropes/

The Global Shipping Industry and Blockchain, April 24, 2018
https://climateyou.org/2018/04/24/our-take-destined-for-the-near-future-is-the-global-shipping-industry-blocktrain-by-climateyou-senior-editor-george-ropes/

Americans Need to Drive Less, November 10, 2019
https://climateyou.org/2019/11/10/americans-need-to-drive-less-by-climateyou-senior-editor-george-ropes/

Innovation and Technology

Solar Plane a Success, April 25, 2016
https://climateyou.org/2016/04/25/solar-plane-a-success/

Waste-to-Fuel Revolution, October 28, 2016
https://climateyou.org/2016/10/28/waste-to-fuel-revolution/

Vacuum up that Carbon!, October 29, 2017
https://climateyou.org/2017/10/29/vacuum-up-that-carbon-by-george-ropes-climateyou-senior-editor/

Geoengineering is Not the Answer to Climate Change, May 4, 2018
https://climateyou.org/2018/05/04/our-take-geoengineering-is-not-the-answer-to-climate-change-by-climateyou-senior-editor-george-ropes/

Let's Capture Methane to Delay Climate Change, May 26, 2019
https://climateyou.org/2019/05/26/our-take-lets-capture-methane-to-delay-climate-change-by-climateyou-senior-editor-george-ropes/

Oceanic Methane Hydrates: The Next Energy Source?, January 10, 2020
https://climateyou.org/2020/01/10/our-take-oceanic-methane-hydrates-the-next-energy-source-by-climateyou-senior-editor-george-ropes/

Dimension 2   Weather and Climate

Extreme Weather

Droughts and Floods, July 10, 2015
https://climateyou.org/2015/07/10/droughts-floods/

HOT HOT FEBRUARY BREAKS RECORDS, March 17, 2016
https://climateyou.org/2016/03/17/hot-hot-february-breaks-records/

Global Droughts & Climate Change, April 8, 2016
https://climateyou.org/2016/04/08/global-droughts-climate-change/

Climate Change & What we can do, December 4, 2016
https://climateyou.org/2016/12/04/climate-you-commentary-climate-change-what-we-can-do/

Peak Hurricane Season and Climate Change = Mega Harvey, August 29, 2017
https://climateyou.org/2017/08/29/peak-hurricane-season-climate-change-mega-harvey/

Heat Projections Higher in Recent Climate Models, October 4, 2021
https://climateyou.org/2021/10/04/our-take-heat-projections-higher-in-recent-climate-models/

February 2021: Texas Freeze, Winter Storms, and the Polar Vortex, June 4, 2021
Andrew Dillon Bustin
https://climateyou.org/2021/06/04/february-2021-texas-freeze-winter-storms-and-the-polar-vortex/

Oceans and Coasts

If You Live On or Near the Coast, You're Gonna Get Wet, July 14, 2017
https://climateyou.org/2017/07/14/if-you-live-on-or-near-the-coast-youre-gonna-get-wet/

Harvey Home Owners: Stay or Go, Pay or Pray, Rebuild or Relocate?, September 5, 2017
https://climateyou.org/2017/09/05/our-take-harvey-home-owners-stay-or-go-pay-or-pray-rebuild-or-relocate/

The Sea Level Rise and Coal Connection, October 28, 2017
https://climateyou.org/2017/10/28/our-take/

Utqiagvik, Alaska, Ground Zero for Climate Change, December 19, 2017
https://climateyou.org/2017/12/19/utiagvik-alaska-ground-zero-for-climate-change-by-climateyou-senior-editor-george-ropes/

Time to Sell Your Beachfront Property?, February 16, 2018
https://climateyou.org/2018/02/16/time-to-sell-your-beachfront-property-by-senior-editor-and-contributor-george-ropes/

Stolen Beaches to Pave Paradise, July 3, 2018
https://climateyou.org/2018/07/03/our-take-stolen-beaches-to-pave-
paradise-by-climateyou-senior-editor-george-ropes/

Warmer Waters & Climate Change Means Bad Weather, July 25, 2018
https://climateyou.org/2018/07/25/warmer-waters-climate-change-means-
bad-weather-by-climateyou-senior-editor-george-ropes/

Dimension 3   Consequences for Nature and People

Deforestation, Fires, and Species Extinction

Deforestation, Climate Change: There Is Still No Planet B, August 1, 2017
https://climateyou.org/2017/08/01/our-take-deforestation-climate-change-
there-still-is-no-planet-b/

Forests Laying Down on the Carbon Job, October 18, 2017
https://climateyou.org/2017/10/18/our-take-forests-laying-down-on-
the-carbon-job/

Climate Change Pushing Species Extinction, March 1, 2018
https://climateyou.org/2018/03/01/climate-change-pushing-species-
extinction-by-climateyou-senior-editor-george-ropes/

Amazon fires won't mean the end of all life on Earth, August 30, 2019
https://climateyou.org/2019/08/30/amazon-fires-wont-mean-the-end-
of-all-life-on-earth-by-climateyou-senior-editor-george-ropes/

Embers of Australian Wildfires 2019–2020, May 26, 2021
Alice Turnbull
https://climateyou.org/2021/05/26/embers-of-australian-wildfires-
2019-2020/

Fanning the Flames: Climate Change and Wildfires in California, June
25, 2021
https://climateyou.org/2021/06/25/fanning-the-flames-climate-change-
and-wildfires-in-california/

Famine and Food Security

Will Agriculture be snubbed at COP21?, November 25, 2015
https://climateyou.org/2015/11/25/will-agriculture-be-snubbed-at-cop21/

Harvesting Carbon as well as Crops, August 12, 2016
https://climateyou.org/2016/08/12/harvesting-carbon-as-well-as-crops/

Urban Agriculture a Key Element in Climate Change, May 31, 2018
https://climateyou.org/2018/05/31/our-take-urban-agriculture-a-key-element-in-climate-change-by-climateyou-senior-editor-george-ropes/

Climate Change at Our Dinner Table (and in Our Trash Bins), June 4, 2021
https://climateyou.org/2021/06/04/climate-change-at-our-dinner-table-and-in-our-trash-bins/

Cities and Power Outages

Increased Flood Risks in Urban Cities, July 29, 2015
https://climateyou.org/2015/07/29/increased-flood-risks-in-urban-cities/

NYC MAYOR De Blasio Pours $100 Million into Downtown Flood Prevention, September 10, 2015
https://climateyou.org/2015/09/10/nyc-mayor-de-blasio-pours-100-million-into-downtown-flood-prevention/

Resilience in Urban Communities, March 29, 2016
https://climateyou.org/2016/03/29/resilience-in-urban-communities/

Climate Change & Power Outages: Update, Puerto Rico, September 23, 2017
https://climateyou.org/2017/09/23/climate-change-power-outages-update-puerto-rico/

Puerto Rico Still Without Power, September 26, 2017
https://climateyou.org/2017/09/26/puerto-rico-still-without-power/

Georgia Power Outages Almost Fixed, September 25, 2017
https://climateyou.org/2017/09/25/georgia-power-outages-almost-fixed/

If New York Goes Under Water, We All Do Too, April 16, 2021
https://climateyou.org/2021/04/16/our-take-if-new-york-goes-under-water-we-all-do-too/

Migration, Conflict, and Population

The New Wave of Climate Refugees from Puerto Rico, October 3, 2017
https://climateyou.org/2017/10/03/our-take-the-new-wave-of-climate-refugees-from-puerto-rico/

Climate Change Driving Immigration, June 21, 2018
https://climateyou.org/2018/06/21/our-take-climate-change-driving-immigration-by-climateyou-senior-editor-george-ropes/

Climate Change and the World Population, February 20, 2019
https://climateyou.org/2019/02/20/climate-change-and-the-world-population-by-climateyou-senior-editor-george-ropes/

Be Prepared for U.S. Climate Migrations, February 8, 2020
https://climateyou.org/2020/02/08/be-prepared-for-u-s-climate-migrattions-by-climateyou-senior-editor-george-ropes/

America's 'Four Great Migrations' & the World They will Transform, February 25, 2021
https://climateyou.org/2021/02/25/americas-four-great-migrations-the-world-they-will-transform/

Dimension 4   Laws and Leaders

International Action

The Green Pontiff Urges Climate Change Initiatives, June 25, 2015
https://climateyou.org/2015/06/25/the-green-pontiff-urges-climate-change-initiatives/

Common Future Moves Forward on Climate Change, July 23, 2015
https://climateyou.org/2015/07/23/common-future-moves-forward-on-climate-change/

World Wins, HFCs Lose, October 23, 2016
https://climateyou.org/2016/10/23/hold-the-gas/

Nuclear Autumn Means Climate Change Disaster, July 30, 2017
https://climateyou.org/2017/07/30/our-take-nuclear-autumn-means-climate-change-disaster/

Pope's July Climate Meeting Lends Hope for Action, June 29, 2018
https://climateyou.org/2018/06/29/our-take-popes-july-climate-meeting-lends-hope-for-action-by-climateyou-senior-editor-george-ropes/

They Did It! Historic Paris Agreement Reached on Climate Change, December 12, 2015, April 21, 2021
https://climateyou.org/2021/04/21/they-did-it-historic-paris-agreement-reached-on-climate-change-december-12-2015/

Hope for COP26: Increasing Ambition, June 19, 2021
https://climateyou.org/?s=Hope+for+COP26%3A+Increasing+Ambition

Ups and Downs: The Role of the United States

Obama's Clean Power Plan, August 9, 2015
https://climateyou.org/2015/08/09/obamas-clean-power-plan/

Climate Change Goes Forward, with or without the U.S., February 8, 2017
https://climateyou.org/2017/02/08/climate-change-goes-forward-with-or-without-the-u-s/

Deb Haaland Confirmation a Fight Over Fossil Fuels & Climate Future, February 10, 2021
https://climateyou.org/2021/02/10/confirm-debra-haaland-save-us-all/

Protect Labor Rights, Protect our Climate and More, March 11, 2021
https://climateyou.org/2021/03/11/our-take-protect-labor-rights-protect-our-climate-and-more/

Front and Center: President Biden's Climate Plan, June 21, 2021
https://climateyou.org/2021/06/21/front-and-center-president-bidens-climate-plan/

China, India, Europe, and the Oil Producers

China's Xi Jinping, Marxism, and Climate Change, May 17, 2018
https://climateyou.org/2021/04/21/chinas-xi-jinping-marxism-and-climate-change-may-17-2018/

Future Politics to Impact India's Huge Carbon Footprint, July 2, 2018
https://climateyou.org/2018/07/02/our-take-future-politics-of-indias-immense-carbon-footprint-by-climateyou-senior-editor-george-ropes/

China's Population Push Means A Bigger Carbon Footprint, August 14, 2018
https://climateyou.org/2018/08/14/our-take-chinas-population-push-means-a-bigger-carbon-footprint-by-climateyou-senior-editor-george-ropes/

Getting to Net Zero — Can we do it?, June 18, 2019
https://climateyou.org/2019/06/18/getting-to-net-zero-can-we-do-it-by-climateyou-senior-editor-george-ropes/

What Will Happen if the US Quits the Middle East?, January 13, 2020
https://climateyou.org/2020/01/13/our-take-the-us-needs-to-pull-out-of-the-middle-east-by-climateyou-senior-editor-george-ropes/

France Judged Guilty in Climate Case, February 7, 2021
https://climateyou.org/2021/02/07/france-judged-guilty-in-climate-case/

Feeding India's Hunger for Oil Could Doom the Human Race, February 17, 2021
https://climateyou.org/2021/02/17/feeding-indias-hunger-for-oil-could-doom-human-race/

Can China & America Mobilize Together to Stop Climate Change?, March 29, 2021
https://climateyou.org/2021/03/29/our-take-can-china-america-mobilize-together-to-stop-climate-change/

Getting to Net Zero, June 3, 2021
https://climateyou.org/2021/06/03/our-take-getting-to-net-zero/

Heroes

Al Gore at 73: Climate Statesman, June 7, 2021
https://climateyou.org/2021/06/07/our-take-al-gore-at-73-climate-statesman/

Climate Activist Greta Thunberg 'Connects the Dots.', June 7, 2021
https://climateyou.org/2021/06/07/our-take-climate-activist-greta-thunberg-connects-the-dots/

Jim Hansen, Climate Alarm, June 19, 2021
https://climateyou.org/2021/06/19/jim-hansen-climate-alarm/

Dimension 5   Finance

NYC Hotels Take up the Carbon Challenge, December 28, 2015
https://climateyou.org/2015/12/28/nyc-hotels-take-up-the-carbon-challenge-2/

Divesting in Fossil Fuels Shapes the Financial Market, February 17, 2016
https://climateyou.org/2016/02/17/divesting-in-fossil-fuels-shapes-the-financial-market/

Whats a Carbon Bubble?, January 2, 2017
https://climateyou.org/2017/01/02/whats-a-carbon-bubble/

Playing Both Sides of the Fossil Fuel Fence, July 6, 2017
https://climateyou.org/2017/07/06/our-take-playing-both-sides-of-the-fossil-fuel-fence/

Can the Pope Work a Miracle with Big Oil & Big Money?, June 6, 2018
https://climateyou.org/2018/06/06/our-take-can-the-pope-work-a-miracle-with-big-oil-big-money-by-climateyou-senior-editor-george-ropes/

Exxon Held to Account for Climate Denial, March 24, 2019
https://climateyou.org/2019/03/24/exxon-to-account-for-climate-denial-by-climateyou-senior-editor-george-ropes/

The World's Growth Model is Unsustainable, April 10, 2019
https://climateyou.org/2019/04/10/our-take-the-worlds-growth-model-is-unsustainable-by-climateyou-senior-editor-george-ropes/

Norway's Sheer Folly to Invest in Oil and Gas, August 25, 2018, April 25, 2021
https://climateyou.org/2021/04/25/norways-sheer-folly-to-invest-in-oil-and-gas-august-25-2018/

Funds that Can Put Your Investments on a Low-Carbon Diet, April 25, 2021
https://climateyou.org/2021/04/25/funds-that-can-put-your-investments-on-a-low-carbon-diet/

Update: Investor Summits on Climate Risk: Advancing the Clean Trillion, October 6, 2021
https://climateyou.org/2021/10/06/update-investor-summits-on-climate-risk-advancing-the-clean-trillion/

Appendix 2

International and U.S. Timeline Tables

Table 1. From Copenhagen to Paris: United Nations Framework Convention on Climate Change Conferences of the Parties (COPs), 2009 to 2015

COP	Year	Goals and Outcomes
COP 15 Copenhagen, Denmark	**2009**	**Goal**: Limit maximum global average temperature increase to no more than 2° C (3.6° F) above pre-industrial levels. **Outcomes**: No agreement of all Parties.
COP 16 Cancun, Mexico	**2010**	**Goals**: Encourage all countries to reduce greenhouse gas (GHG) emissions and help developing nations deal with climate change. **Outcomes**: • Establishment of Technology Mechanism; • Establishment of Green Climate Fund to provide support to developing countries to assist in mitigating climate change and adapting to its impacts.
COP 17 Durban, South Africa	**2011**	**Goal**: Advance toward global agreement. **Outcomes**: • Roadmap to build and preserve trust among countries;

(Continued)

Table 1.   *(Continued)*

COP	Year	Goals and Outcomes
		• Commitment to deliver new and universal GHG reduction protocol with legal force by 2015 for period beyond 2020.
COP 18 Doha, Qatar	2012	**Goal**: Advance toward global agreement. **Outcomes**: • Timetable to adopt universal climate agreement by 2015, to come into effect in 2020; • New mechanisms on finance, review, adaptation, and technology; voluntary emission pledges from 94 countries; • Establishment of new mechanism to address "Loss and Damage" from extreme weather and from slow-onset climate impacts such as sea level rise in developing countries.
COP 19 Warsaw, Poland	2013	**Goal**: Advance global agreement. **Outcomes**: • Warsaw International Mechanism for "Loss and Damage" associated with climate change; • Flexibility in addressing responsibility of GHG emissions reductions between developing and developed countries; • Formal adoption of United Nations Reducing Emissions from Deforestation and Forest Degradation (REDD+) Program, aimed at preserving world's forests.
COP 20 Lima, Peru	2014	**Goal**: Advance global agreement. **Outcomes**: • Establishment of ground rules on how countries can submit Nationally Determined Contributions (NDCs); • Pledges by developed and developing countries that took capitalization of new Green Climate Fund (GCF) past initial $10 billion target.

Table 1. (*Continued*)

COP	Year	Goals and Outcomes
COP 21 **Paris, France**	**2015**	**Goal**: Negotiate global agreement to limit global warming to well below 2°, preferably to 1.5° C, (2.7° to 3.6° F) compared to pre-industrial levels. **Outcomes**: • Establishment of landmark Paris Agreement, legally binding international treaty on climate change to regularly set national targets, adopted by 196 countries;[1] • Provision of framework for financial, technical, and capacity-building support to those countries who need it; • Reaffirmation that developed countries should take lead in providing financial assistance to countries less endowed and more vulnerable.

Table 2. **From Clinton to Biden: U.S. Presidents and Climate Change, 1993 to 2022**

President	Term	Party	Climate Change Actions
William J. Clinton	1993–2001	Democrat	**1993**: **Climate Change Action Plan** **GHG Target**: Reduce U.S. greenhouse gas emissions to 1990 levels by 2000 **Binding/Non-Binding**: Voluntary **Components**: Education on climate change; mandates for household appliances and buildings to be energy-efficient; increased funding for Green Lights Program **1996**: 4.3 cent per gallon tax on gasoline and transportation fuel

(*Continued*)

---

[1]While legally binding internationally, the Paris Agreement relies on informal compliance processes. https://www.weforum.org/ accessed June 4, 2022

Table 2. (*Continued*)

President	Term	Party	Climate Change Actions
			Political Climate: House and Senate in Republican control staunchly against climate change action
George W. Bush	2001–2009	Republican	**2001**: Refusal to implement the 1997 Kyoto Protocol international treaty, requiring nations to reduce GHG emissions **2002: Climate Change Technology Program** (CCTP) **Components**: $6.3 billion, five-year tax incentive package to stimulate energy efficient technologies in buildings, industrial processes, vehicles, and power generation **Binding/Non-Binding**: Voluntary **Political Climate**: Bush White House claimed that ratifying the Kyoto Protocol would create economic setbacks in the U.S.
Barack Obama	2009–2017	Democrat	**2009: American Clean Energy and Security Act of 2002 (ACES)** Passed in House, never went to Senate; new auto regulations to boost efficiency of cars and light trucks by 2016 **2013:** Climate Action Plan to cut domestic carbon pollution, prepare the U.S. for climate change impacts, and lead international efforts to address global climate change

Table 2. (*Continued*)

President	Term	Party	Climate Change Actions
			2015: Clean Power Plan Initiative **GHG Emission Target**: Reduce carbon dioxide emissions from electrical power generation by 32% by 2030, relative to 2005 levels **Components**: Reduction in emissions from coal-burning power plants and increase in use of renewable energy **2016**: U.S. formally enters the UNFCCC Paris Agreement **Political Climate**: Senate in Republican control; continued opposition to climate change action
Donald Trump	2017–2021	Republican	**2017: Elimination of President Obama's Climate Action Plan** Executive Order to officially nullify President Obama's Clean Power Plan in effort to revive coal industry and replace it with weaker regulations that limit GHG emissions **2017**: U.S. stopped participating in 2015 Paris Agreement, contending agreement would "undermine" U.S. economy **2017–2020**: Lifted bans on oil and gas exploration in Arctic National Wildlife Refuge and parts of National Petroleum Reserve in Alaska, coastal waters and areas

(*Continued*)

Table 2.   *(Continued)*

President	Term	Party	Climate Change Actions
			formerly protected as National Monuments in Utah **2018: Affordable Clean Energy Rule** Clean Power Plan rolled back by EPA; erased requirements to reduce emissions from power plants by 32% below 2005 levels by 2030 **GHG Emissions Target**: No national target **Political Climate**: Republican control of House and Senate; continued opposition to climate change action
Joseph R. Biden 2021	2021– present	Democrat	**January 2021**: **Re-instatment of President Obama's Climate Action Plan** **February 2021**: U.S. re-enters UNFCCC Paris Agreement **April 2021**: High-level International *Leaders Climate Ambition Summit* with 40 Heads of State on Earth Day **Executive Orders:** Creation of *Special Presidential Envoy for Climate* to have seat on National Security Council; Establishment of *National Climate Task Force*, assembling leaders from 21 federal agencies and departments to enable whole-of-government approach to combatting climate crisis;

Table 2.   (*Continued*)

President	Term	Party	Climate Change Actions
			Establishment of *Civilian Climate Corps; Revitalize Energy Communities: Interagency Working Group on Coal and Power Plant Communities and Economic Revitalization*
			Legislation **U.S. Infrastructure Bill** Signed November 2021 with $50 billion to protect country against climate change impacts; bipartisan bill also funded hydrogen hubs and EV charging stations
			CHIPS Act Signed in August, 2022; boost for U.S. semiconductor production included many clean energy provisions
			Inflation Reduction Act Major climate change legislation passed in August 2022; massive incentives for U.S. renewable energy production and use and subsidies for purchases of electric vehicles; many environmental justice provisions **GHG Emissions Target:** Reduce U.S. emissions by 50% of 2005 levels by 2030 **Political Climate:** Democratic control of House of Representatives; narrow control of Senate, but able to make deal with Senator Joe Manchin to pass historic climate bill

Appendix 3

Climate Change Organizations Throughout the World

Groups are mobilizing to advance action on climate change across the globe. Appendix 3 highlights some of the many organizations that are working to solve the climate crisis. Each section of Appendix 3 presents climate change organizations in different countries or continents, with sections on United States, Canada, United Kingdom, Europe, Australia and Oceania, Asia, Africa, and Latin America. For each climate change organization, we list the website, mission statement, date of founding, and location, if available. Please look for your country or continent and visit sites of interest for more information.

3A. UNITED STATES

350.org
www.350.org

350.org is an international movement of ordinary people working to end the age of fossil fuels and build a world of community-led renewable energy for all. Founded in 2008. Location: Boston, Massachusetts.

American Association of State Climatologists
www.stateclimate.org

The American Association of State Climatologists is committed to advancing the development and delivery of science-based climate services on a local and state level. Founded in 1976. Location: Asheville, North Carolina

Carbonfund.org

www.carbonfund.org

The Carbon Fund is leading the fight against climate change, making it easy and affordable for any individual, business, or organization to reduce and offset their climate impact, and hasten the transition to a clean energy future. Location: East Aurora, New York.

Center for Climate and Energy Solutions

www.c2es.org

The mission of the Center for Climate and Energy Solutions is to advance strong policy and action to reduce greenhouse gas emissions, promote clean energy, and strengthen resilience to climate impacts. A key objective is a national market-based program to reduce emissions cost-effectively. A sound climate strategy is essential to ensure a strong, sustainable economy. Founded in 2011. Location: Arlington, Virginia.

Center for Climate and Life

www.climateandlife.columbia.edu

The mission of the Center for Climate and Life is to deepen understanding of how climate affects human sustainability by accelerating innovative climate research at Columbia University and providing expert analysis on climate impacts and solutions. Founded in 2016. Location: New York, New York.

Chesapeake Climate Action Network

www.chesapeakeclimate.org/mission

The Chesapeake Climate Action Network (CCAN) is the first grassroots, non-profit organization dedicated exclusively to fighting global warming in Maryland, Virginia, Washington, DC, and nationwide. Its mission is to build a diverse movement powerful enough to put the region on the path to climate stability, while using proximity to the nation's capital to inspire action in neighboring states, regions nationwide, and countries around the world. Founded in 2002. Location: States surrounding Chesapeake Bay.

Citizens' Climate Lobby

www.citizensclimatelobby.org

Citizens' Climate Lobby is a non-profit, non-partisan, grassroots advocacy climate change organization focused on national policies to address climate change. Founded in 2007. Location: Coronado, California.

Citizens for Pennsylvania's Future

www.pennfuture.org

PennFuture is leading the transition to a clean energy economy in Pennsylvania and beyond. Citizens for Pennsylvania's Future is protecting the state's air, water, and land, and empowering its citizens to build sustainable communities for future generations. Founded in 1998. Location: Harrisburg, Pennsylvania.

Climate Central

www.climatecentral.org

Climate Central is an independent organization of leading scientists and journalists researching and reporting the facts about our changing climate and its impact on the public. Founded in 2008. Location: Princeton, New Jersey.

Climate Emergency Fund

www.climateemergencyfund.org

The Climate Emergency Fund (CEF) is mainstreaming the climate crisis; defining the climate agenda as a non-partisan issue; supporting new, diverse groups and activists; and bringing together coalitions to increase the spread of climate messaging. Location: Beverly Hills, California.

Climate Hawks Vote

The mission of Climate Hawks Vote is to identify, train, and elect individual climate hawk leaders, while generating a political environment in which those leaders have the power to advance policies needed to address climate change. Founded in 2013. Location: California.

Climate Justice Alliance

www.climatejusticealliance.org

The Climate Justice Alliance (CJA) was formed to create a new center of gravity in the climate movement by uniting frontline communities and organizations into a formidable force. The translocal organizing strategy and mobilizing capacity are building a Just Transition away from extractive systems of production, consumption, and political oppression, and towards resilient, regenerative, and equitable economies. The process of transition must place race, gender, and class at the center of the climate change solutions equation in order to make it a truly Just Transition. Founded in 2013.

Climate Leadership Council

www.clcouncil.org

The Climate Leadership Council represents the broadest climate coalition in U.S. history. It promotes a bipartisan carbon dividends framework as the most politically viable, equitable, and pro-business climate solution. Founded in 2017. Location: Washington, DC.

Climate One

www.climateone.org

Climate One is a special project of the Commonwealth Club of California. It is a public forum and podcast series for conversations on climate change and its implications for society, energy systems, economy, and the natural environment. Founded in 2007. Location: San Francisco, California.

Climate Savers Computing Initiative

www.climatesaverscomputing.org

The Climate Savers Computing Initiative was a non-profit group of consumers, businesses, and conservation organizations dedicated to promoting smart technologies that improve power efficiency and reduce energy consumption of computers. In July 2012, Climate Savers Computing Initiative combined with the Green Grid, and its programs continue within that organization. Founded in 2007. Location: Portland, Oregon.

ClimateYou Alliance

www.climateyou.org

The ClimateYou Alliance informs the public about the complex climate system that envelops us, emphasizing the changes induced by the burning of fossil fuels, and the measures needed to mitigate or adapt to those changes. It empowers people to act to ensure that the Earth remains a hospitable and sustainable environment for humans and all forms of life. Through its educational programs, ClimateYou provides students with opportunities to learn, write, and voice their views about climate change. Founded in 2008. Location: Tarrytown, New York.

Coastal Risk Consulting

Coastal Risk Consulting, LLC, is an American startup climate adaptation technology and consulting company. Coastal Risk provides individuals, businesses, and local governments with climate impact modeling technology, available as an online software-as-a-service (SaaS), that allows property owners to assess their vulnerability to flooding related to sea level rise and climate change impacts, and assists in adaptation and resiliency decision-making. Founded in 2014. Location: Plantation, Florida.

Coltura

www.coltura.org

Coltura is an American environmental activist group that promotes environmental policies and produces cultural works aimed at ending America's use of gasoline, and advocates for policies to phase out the sales of new gasoline-powered vehicles by 2030. Founded in 2014. Location: Seattle, Washington.

CoolCalifornia.org

www.coolcalifornia.arb.ca.gov

The mission of CoolCalifornia.org is to provide all Californians with the tools they need to take action to protect the climate and keep California cool. Location: California.

Earthwatch Institute
www.earthwatch.org

Earthwatch connects people with scientists worldwide to conduct environmental research and empowers them with the knowledge they need to conserve the planet. Founded in 1971. Location: Boston, Massachusetts.

Ecology Center — Ann Arbor
www.ecocenter.org

The Ecology Center is a membership-based non-profit environmental organization. It works at the local, state, and national levels on environmental justice, health, waste, and community issues. It was formed after the first Earth Day in 1970 by community activists. Since its founding, it has run demonstrations and campaigns to promote recycling, health care, education, and awareness about healthy foods and products. Founded in 1970. Location: Ann Arbor, Michigan.

Ecology Center — Berkeley
www.ecologycenter.org

The Ecology Center is a non-profit organization that works to provide environmental education and reduce the ecological footprint of urban residents. Founded in 1969. Location: Berkeley, California.

Ecosystem Marketplace
www.ecosystemmarketplace.com

Ecosystem Marketplace, an initiative of the non-profit organization Forest Trends, is a leading global source of information on environmental finance, markets, and payments for ecosystem services. Founded in 2004. Location: Washington, DC.

Environment California
www.environmentcalifornia.org

Environment California's mission is to transform the power of our imaginations and our ideas into change that makes our world a greener and healthier place for all. Founded in 1970. Location: Los Angeles, California.

Environmental and Energy Study Institute

www.eesi.org

Founded by a bipartisan group of members of Congress to inform the debate and decision-making on energy and environmental policies, the Environmental and Energy Study Institute (EESI) is a 501(c)(3) non-profit organization dedicated to promoting sustainable societies. Today, its mission is to advance science-based solutions for climate change, energy, and environmental challenges in order to achieve the vision of a sustainable, resilient, and equitable world. Founded in 1984. Location: Washington, DC.

ETC Group — Eco-Justice

ETC Group is an international organization based in the U.S. dedicated to "the conservation and sustainable advancement of cultural and ecological diversity and human rights." The full legal name is Action Group on Erosion, Technology and Concentration. "ETC" is intended to be pronounced "*et cetera.*" The ETC Group publishes opinions on scientific research; its staff and board members come from a variety of backgrounds, including community and regional planning, ecology and evolutionary biology, and political science.

Four Twenty Seven

www.427mt.com

Four Twenty Seven, an affiliate of Moody's, is a leading publisher and provider of data, market intelligence, and analysis related to physical climate and environmental risks. The mission is to catalyze climate adaptation and resilience investments by enabling the integration of climate science into business and policy decisions. Founded in 2012. Location: Berkeley, California.

Great March for Climate Action

www.climatemarch.org

The goal of the Great March for Climate Action is to change the hearts and minds of Americans, elected leaders, and people across the world to act now to address the climate crisis. Founded in 2013. Location: Des Moines, Iowa.

Green Corps

www.greencorps.org

The mission of Green Corps is to train organizers, provide field support for today's critical environmental campaigns, and graduate activists who possess the skills, temperament, and commitment to fight and win tomorrow's environmental battles. Founded in 1992.

Green Light New Orleans

www.greenlightneworleans.org

Green Light New Orleans provides sustainable solutions to individual homes and encourages collective action to create a more resilient community. What began as one man's "light bulb moment," Green Light New Orleans now operates one of the largest energy efficiency programs in New Orleans. Founded in 2006. Location: New Orleans, Louisiana.

Greenbelt Alliance

www.greenbelt.org

The mission of the Greenbelt Alliance is to educate, advocate, and collaborate to ensure the Bay Area's lands and communities are resilient to a changing climate. Founded in 1958. Location: San Francisco, California.

Greenpeace USA

www.greenpeace.org/usa

Greenpeace USA is the U.S. affiliate of Greenpeace, an international environmental non-profit organization. Founded in 1975. Location: Washington, DC.

Honor the Earth

www.honorearth.org

The mission of Honor the Earth is to create awareness and support for Native environmental issues and to develop needed financial and political resources for the survival of sustainable Native communities. Honor the Earth develops these resources by using music, the arts, the media, and Indigenous wisdom to ask people to recognize our joint dependency on the Earth and be a voice for those not heard. Founded in 1993. Location: Callaway, Minnesota.

Information Technology and Innovation Foundation
www.itif.org/issues/clean-energy-innovation

The Information Technology and Innovation Foundation (ITIF) is an independent 501(c)(3) non-profit, non-partisan research and educational institute — a think tank. Its mission is to formulate, evaluate, and promote policy solutions that accelerate innovation and boost productivity to spur growth, opportunity, and progress. ITIF's goal is to provide policy makers around the world with high-quality information, analysis, and recommendations they can trust. To that end, ITIF adheres to a high standard of research integrity with an internal code of ethics grounded in analytical rigor, policy pragmatism, and independence from external direction or bias. Founded in 2006. Location: Washington, DC.

InsideClimate News
www.insideclimatenews.org

InsideClimate News is a non-profit news organization, focusing on environmental journalism. The publication writes that it "covers clean energy, carbon energy, nuclear energy and environmental science — plus the territory in between where law, policy and public opinion are shaped." Founded in 2007. Location: Brooklyn, New York.

Mobilize Earth
www.mobilize.earth

Mobilize Earth is setting out on a new course to mitigate the climate and ecological emergency by bringing together people from all walks of life. The organization is working to mobilize everyone — industry, activists, politicians, farmers, immigrants — to respond to the climate and ecological emergency. Founded in 2020. Location: Edgewater, New Jersey.

Montana Environmental Information Center
www.meic.org

The Montana Environmental Information Center was founded by Montanans to protect and restore Montana's natural environment. It functions as a non-profit environmental advocacy group. Founded in 1973. Location: Helena, Montana.

National Center for Science Education
www.ncse.ngo

The National Center for Science Education (NCSE) is a not-for-profit membership organization in the U.S. whose mission is to educate the press and the public on the scientific and educational aspects of controversies surrounding the teaching of evolution and climate change, and to provide information and resources to schools, parents, and other citizens working to keep those topics in public school science education. Founded in 1981. Location: Oakland, California.

NextGen America
www.nextgenamerica.org

NextGen America is a progressive advocacy non-profit and political action committee created by Tom Steyer. It registers voters, encourages voting, and helps to inform young voters on issues including the climate. Founded in 2013. Location: San Francisco, California.

One Tree Planted
www.onetreeplanted.org

As an environmental charity, the organization One Tree Planted is on a mission to make it simple for anyone to help the environment by planting trees. One dollar, one tree. Founded in 2014. Location: Shelburne, Vermont.

Pacific Environment
www.pacificenvironment.org

The organization Pacific Environment nurtures the courage and creativity of local leaders to find new ways to protect all of us from climate breakdown, air pollution, water loss, and plastic waste. The organization advocates for strong protections for people and the planet at the highest international levels of government. Founded in 1987. Location: San Francisco, California.

People's Climate Movement

www.peoplesclimate.org

The People's Climate Movement (PCM) is a coalition of civil society organizations, including environmental and religious organizations, trade unions and social justice groups, in the U.S. that advocates for political and social change to reverse or mitigate the effects of climate change. PCM emphasizes the inclusion of underrepresented groups, job creation, and economic prosperity. Since 2014, PCM has organized marches in the U.S. to raise awareness and demand action on climate issues.

Physicians for Social Responsibility

www.psr.org

Physicians for Social Responsibility (PSR) is a physician-led organization in the U.S. working to protect the public from the threats of nuclear proliferation, climate change, and environmental toxins. It produces and disseminates publications, provides specialized training, offers written and oral testimony to congress, conducts media interviews, and delivers professional and public education. Founded in 1961. Location: Boston, Massachusetts.

Power Shift Network

www.powershift.org

The Power Shift Network mobilizes the collective power of young people to mitigate climate change and create a just, clean energy future and resilient, thriving communities for all. Founded in 2004. Location: Washington, DC.

Rainforest Foundation US

www.rainforestfoundation.org

The mission of the Rainforest Foundation US is to protect rainforests in partnership with indigenous peoples. Founded in 1989. Location: New York, New York.

Rising Tide North America

www.risingtidenorthamerica.org

Rising Tide North America is a grassroots network of groups and individuals in North America organizing action against the root causes of climate change and work towards a non-carbon society. Founded in 2000.
Location: Philadelphia, Pennsylvania.

Rocky Mountain Institute

The Rocky Mountain Institute (RMI) is dedicated to research, publication, consulting, and lecturing in the general field of sustainability, with a special focus on profitable innovations for energy and resource efficiency. Founded in 1982. Location: Basalt, Colorado.

Sierra Club

www.sierraclub.org

The Sierra Club is an enduring and influential grassroots environmental organization in the U.S. The organization amplifies the power of its 3.8 million members and supporters to defend everyone's right to a healthy world. Founded in 1892.
Location: Oakland, California.

Sierra Nevada Alliance

www.sierranevadaalliance.org

The Sierra Nevada Alliance is a network of conservation groups encompassing 24 watersheds of the 650-kilometer-long Sierra Nevada in California and Nevada. Beginning in 1993, the Alliance protects and restores Sierra Nevada lands, watersheds, wildlife, and communities. Founded in 1993.
Location: South Lake Tahoe, California.

Sierra Student Coalition

www.sierraclub.org/youth

The Sierra Student Coalition (SSC) is a network of young people, aged 14 to 35, organizing for climate, racial, and economic justice. It offers training programs, popular education, leadership development opportunities, campaigns,

resources, and a community of support for youth across the country to take action on the issues that they care most about. Founded in 1991.
Location: Washington, DC.

Solar Cookers International

www.solarcookers.org

Solar Cookers International (SCI) improves human and environmental health by supporting the expansion of effective carbon-free solar cooking in world regions of greatest need. SCI leads through advocacy, research, and strengthening the capacity of the global solar cooking movement. Founded in 1987.
Location: Sacramento, California.

Sunrise Movement

www.sunrisemovement.org

The Sunrise Movement is a youth movement to stop climate change and create millions of good jobs in the process. The organization is building an army of young people to make climate change an urgent priority across America, end the corrupting influence of fossil fuel executives on our politics, and elect leaders who stand up for the health and wellbeing of all people. Founded in 2017. Location: Washington, DC.

SustainUS

www.sustainus.org

SustainUS is a non-profit, non-partisan, youth-led advocacy group in the U.S. Its goal is to improve youth participation and youth empowerment as it relates to advancing sustainable development. SustainUS works particularly with youth and young people aged 13–29 and on United Nations conferences related to youth and/or sustainability. Founded in 2001. Location: Washington, DC.

TerraPass

www.terrapass.com

The mission of TerraPass is to provide the resources necessary for companies and individuals to understand and take responsibility for their climate impact. Founded in 2004. Location: San Francisco, California.

The Climate Group
www.theclimategroup.org

The Climate Group's mission is driving climate action. Science says that we must cap global warming at 1.5° C (2.7° F) to avoid the disastrous effects of climate change. To have a fighting chance at doing this, we must halve global emissions in the next 10 years — that is why the 2020s have to be the Climate Decade. Founded in 2004. Location: New York and London.

The Climate Mobilization
www.theclimatemobilization.org

The Climate Mobilization organization is building a movement of people across the U.S. to reclaim our future by initiating an emergency-speed, whole-society Climate Mobilization, reversing global warming and restoring a safe climate. Founded in 2014. Location: New York, New York.

The Climate Reality Project
www.climaterealityproject.org

The mission of this organization, the Climate Reality Project, is to catalyze a global solution to the climate crisis by making urgent action a necessity across every sector of society. Founded in 2006. Location: Washington, DC.

Union of Concerned Scientists
www.ucsusa.org

The mission of Union of Concerned Scientists (UCS) is to use rigorous, independent science to solve our planet's most pressing problems. Joining with people across the country, the people in UCS combine technical analysis and effective advocacy to create innovative, practical solutions for a healthy, safe, and sustainable future. Founded in 1969. Location: Cambridge, Massachusetts.

World War Zero
www.worldwarzero.com

World War Zero is an American coalition launched by John Kerry in 2019 to fight the climate crisis. The main goal of the coalition is to hold more than 10 million "climate conversations" with citizens across the political spectrum. Founded in 2019.

World Wildlife Foundation U.S.

www.worldwildlife.org

The mission of World Wildlife Foundation U.S. is to conserve nature and reduce the most pressing threats to the diversity of life on Earth. Founded in 1961. Location: Washington, DC.

Yale Program on Climate Change Communication

www.climatecommunication.yale.edu

The Yale Program on Climate Change Communication (YPCCC) is a research center within the Yale School of Forestry & Environmental Studies that conducts scientific research on public climate change knowledge, attitudes, policy preferences, and behavior at the global, national, and local scales. Founded in 2005. Location: New Haven, Connecticut.

Young Voices for the Planet

www.youngvoicesfortheplanet.com

The mission of the Young Voices for the Planet film series is to limit the magnitude of climate change and its impacts by empowering children and youth, through uplifting and inspiring success stories, to take an essential role in informing their communities — and society at large, challenging decision-makers, and catalyzing change. The organization documents youth speaking out, creating solutions, and leading the change. Location: Thurmont, Maryland.

3B. CANADA

Canadian Parks and Wilderness Society

www.cpaws.org

The Canadian Parks and Wilderness Society (CPAWS) is Canada's only nationwide charity dedicated solely to the protection of public land and water, and ensuring parks are managed to protect the nature within them now and in the future. In the past 50 years and more, the society has played a lead role in protecting over half a million square kilometers — an area bigger than the entire Yukon Territory. The organization's vision is to protect at least half of Canada's public land and ocean in a framework of reconciliation — for the benefit of both wildlife and humans. Founded in 1963. Location: Ottawa, Ontario.

Climate Action Network — Canada

www.climateactionnetwork.ca

Climate Action Network — Canada acts as a clearinghouse and network for more than 100 Canadian groups working on climate change. CANet's successful and informal listserve provides a forum for instantaneous communication on policy developments and devising cooperative strategies. CANet's key strategy is to use information effectively. Founded in 1989. Location: Ottawa, Ontario.

David Suzuki Foundation

www.davidsuzuki.org

Through evidence-based research, education, and policy analysis, the David Suzuki Foundation works to conserve and protect the natural environment, and help create a sustainable Canada. The foundation regularly collaborates with non-profit and community organizations, all levels of government, businesses, and individuals. Founded in 1990. Location: Vancouver, British Columbia.

Ecotrust Canada

www.ecotrust.ca

EcoTrust Canada works to establish ecologically sustainable communities (especially First Nations), renewable forests and forest practices, and protect fisheries. EcoTrust also funds community development and housing projects. Founded in 1995. Location: Vancouver, British Columbia.

Greenpeace Canada

www.greenpeace.org/international

Greenpeace's early focus was on strengthening the international Climate Change Convention through direct action and lobbying activities. Its subsequent focus has been on stopping any further exploration and development of fossil fuels and launching the "Solar Century." Greenpeace is famous for its use of direct action, a tool it employs selectively. Founded in 1969. Location: Vancouver, British Columbia.

KAIROS (Canadian Ecumenical Justice Initiatives)

www.kairoscanada.org

KAIROS reaches out to Canadian church communities by organizing workshops across the country. In one year alone, between June 2000 and June 2001, more than 4,000 people attended over 100 workshops on "creating a climate for change," promoting energy efficiency in religious buildings and reducing human ecological footprints. Founded in 2001. Location: Toronto, Ontario.

Nature Canada

www.naturecanada.ca

Nature Canada includes a network of 350 naturalist clubs, leveraging their many members to complete research projects and make policy recommendations. They focus on endangered species, bird conservation, water resources, and parks and protected areas. Founded in 1939. Location: Ottawa, Ontario.

Nature Conservancy of Canada

www.natureconservancy.ca/en

The Nature Conservancy of Canada (NCC) funds the Natural Areas Conservation Program, the largest land conservation program in Canada. The NCC works closely with the Federal government. Founded in 1962. Location: Toronto, Ontario.

Pembina Institute

www.pembina.org

The Pembina Institute is a Canadian non-profit think tank focused on energy. The institute aims to reduce the harmful impacts of fossil fuels while supporting the transition to an energy system that is clean and safe. Founded in 1985. Location: Alberta, Canada.

3C. UNITED KINGDOM

British Ecological Society

www.britishecologicalsociety.org

As the world's oldest ecological society, the British Ecological Society aims to ensure that as many people as possible are educated on all types of ecological matters. They have more than 6,400 members around the world, working to advance ecological science. Founded in 1913. Location: London, UK.

Campaign to Protect Rural England

www.cpre.org.uk

Focusing mainly on waste management, transport, and energy, the Campaign to Protect Rural England is a charity that works to protect the UK's countrysides. Founded in 1926. Location: London, UK.

Committee on Climate Change

www.theccc.org.uk

The Climate Change Committee (CCC) is an independent, statutory body established under the Climate Change Act 2008. Its purpose is to advise the UK and devolved governments (separate legislatures and executives in Scotland, Wales, and Northern Ireland) on emissions targets. It reports to Parliament on progress made in reducing greenhouse gas emissions and preparing for and adapting to the impacts of climate change. Founded in 2008.

Energy Saving Trust

www.energysavingtrust.org.uk

The Energy Saving Trust (EST) is a British organization devoted to promoting energy efficiency, energy conservation, and the sustainable use of energy, thereby reducing carbon dioxide emissions and helping to prevent human-caused climate change. It was founded in the UK as a government-sponsored initiative, following the global Earth Summit. Founded in 1992. Location: London, UK

Environmental Law Foundation

www.elflaw.org

This charity, the Environmental Law Foundation, has the Prince of Wales as its president. It helps the voice of ordinary people and communities to be heard on matters affecting the environment in which they live. Founded in 1992. Location: Worcestershire, UK.

Ethical Consumer

www.ethicalconsumer.org

The Ethical Consumer Research Association Ltd. (ECRA) is a British not-for-profit publisher, research, and campaign organization that publishes information on the social, ethical, and environmental behavior of companies and issues around trade justice and ethical consumption. Ethical Consumer strives to promote sustainability in business practices by having a greener approach for their products. The website gives a rating to more than 40,000 products, providing consumers the ability to see how green items are before buying. Founded in 1989. Location: Manchester, UK.

Friends of the Earth

https://friendsoftheearth.uk/

Through campaigns such as reducing air pollution, saving British bees, and other important environmental issues, the vision of the Friends of the Earth is of a peaceful and sustainable world based on societies living in harmony with nature. Member groups include England, Wales, and Northern Ireland. Founded in 1969. Location: London, UK.

Green Alliance

www.green-alliance.org.uk

Green Alliance is an independent think tank and charity focused on ambitious leadership for the environment. They work with influential leaders in business, non-governmental organizations (NGOs), and politics to accelerate political action and create transformative policy for a green and prosperous UK. Founded in 1979. Location: London, UK.

Institution of Environmental Sciences

www.the-ies.org

The Institution of Environmental Sciences (IES) is a charitable organization that promotes public awareness by supporting professional researchers and academics working in this crucial arena. As a seminal science organization, it is consulted by government and other interested parties on environmental issues. The Institution has strong ties with higher education, enhancing environmental science and sustainable development in universities and colleges nationally and internationally. Founded in 1971. Location: London, UK.

Keep Britain Tidy

www.keepbritaintidy.org

Keep Britain Tidy is a UK-based independent environmental charity that campaigns to reduce litter, improve local places, and prevent waste. Partnering with businesses, organizations, and individuals, it works to create clean beaches, parks, and streets and to foster sustainable practices. Founded in 1960. Location: London, UK.

Recycle Now

www.recyclenow.com

Recycle Now is the national recycling campaign for England, supported and funded by Government, managed by the Waste and Resources Action Programme (WRAP), and used locally by over 90% of English authorities. The campaign is here to help people to recycle more things, more often. More than 6 out of 10 of us now describe ourselves as committed recyclers, compared to less than half when the campaign began in 2004. Founded in 2004. Location: Banbury, Oxfordshire, UK.

RenewableUK

www.renewableuk.com

RenewableUK, formerly known as the British Wind Energy Association, is the trade association for wind power, wave power, and tidal power industries in the UK. RenewableUK has over 660 corporate members, from wind, wave, and tidal stream power generation and associated industries. Founded in 1978. Location: London, UK.

The Wildlife Trusts
www.wildlifetrusts.org

With more than 800,000 members spread across close to 50 different groups across the UK, the Wildlife Trusts are responsible for protecting many natural environments. Whether parks, woods, or nature reserves, these trusts fight to keep them protected. Founded in 1912. Location: Newark-on-Trent, Nottinghamshire, UK.

3D. EUROPE

Agir Pour l'Environnement
www.agirpourlenvironnement.org

Agir Pour l'Environnement (Acting for the Environment) is an association of citizen mobilization working for a livable planet. The association puts pressure on politicians and economic decision-makers by leading campaigns bringing together a large network of associations and citizens. In order to keep intact its independence of action, the association refuses any funding from public authorities. Founded in 1997. Location: Paris, France.

Climate Action Network Europe
www.climatenetwork.org

Climate Action Network Europe (CAN) is a leading environmental organization in Europe and envisions a world that actively strives to achieve global climate protection, while still promoting equity, social justice, and sustainable development. The organization, which comprises of more than 170 member organizations, is present in over 38 European nations and represents over 47 million citizens. It mainly advocates for sustainable climate, as well as energy and development policies across Europe and beyond. Founded in 1989. Location: Bonn, Germany.

Climate Analytics

www.climateanalytics.org

Climate Analytics brings cutting-edge science and policy analysis to bear on one of the most pressing global problems of our time — human-induced climate change. As the urgency of this problem has grown, the Climate Analytics team of about 100 counts 32 different nationalities working at offices in Togo, the U.S., Australia, and Germany, as well in a diverse set of countries including the Bahamas, Nepal, Bhutan, Burkina Faso, Samoa, Trinidad and Tobago, and the UK. Founded in 2008. Location: Berlin, Germany.

Coalition Clean Baltic

www.ccb.se

Coalition Clean Baltic (CCB) was established in Helsinki, Finland, when environmental non-governmental organizations (NGOs) from the countries of the Baltic Sea Region became united to cooperate in activities concerning the Baltic Sea environment. Their main objective is to advocate for protection, preservation, and enhancement of the environment and natural resources of the Baltic Sea. Founded in 1990. Location: Uppsala, Sweden.

ECODES

www.ecodes.org

The ECODES (Foundation for Ecology and Development) works towards sustainable and environmentally friendly development, to strengthen dialogue and collaboration with all stakeholders in the implementation of actions and programs that promote sustainable development, and to enable social change. They provide ideas and solutions as well as critical commentary based on professional expertise. Their main areas of focus are social responsibility, climate change resulting from global warming, water management, reducing consumption, and development cooperation. Founded in 1992. Location: Zaragoza, Spain.

European Cyclists' Federation

www.ecf.com

Its slogan is, "Bicycling is the European way!" Founded by just 12 bicycle user associations, the European Cyclists' Federation (ECF) has grown to become the umbrella federation for all Europe's cycling organizations. Its aim is to promote the use of bikes in urban centers as the main means of transport and to see cycle tourism recognized as a sustainable economic factor, as well as an environmentally friendly mode of transport. Founded in 1983. Location: Brussels, Belgium.

European Environment Agency

www.eea.europa.eu

The European Environment Agency (EEA) is an agency of the European Union, whose task is to provide sound, independent information on the environment. The EEA aims to support sustainable development by helping to achieve significant and measurable improvement in Europe's environment, through the provision of timely, targeted, relevant, and reliable information to policymakers and the public. Founded in 1990. Location: Copenhagen, Denmark.

European Environmental Bureau

www.eeb.org

European Environmental Bureau (EEB) is the largest environmental preservation organization in Europe. It is currently comprised of more than 160 member organizations, in over 35 countries. They stand for sustainable development, environmental justice, and participatory democracy. Founded in 1974. Location: Brussels, Belgium.

Forests and the European Union Resource Network

www.fern.org

Forests and the European Union Resource Network (FERN) is a Dutch-based environmental organization that was formed with the sole aim of protecting forests and the communities that depend on them. Founded in 1995. Location: Brussels, Belgium.

Friends of the Earth Europe

www,friendsoftheearth.eu

The Friends of the Earth Europe is among the largest grassroots environmental organizations in the continent. It has united over 75 national organizations, over 5,000 of community-based groups, and more than two million global supporters. The organization envisions a natural, peaceful, and sustainable environment, where people, animals, and nature can all live in harmony. Founded in 1986. Location: Brussels, Belgium.

International Solar Energy Society

www.ises.org

The members of the International Solar Energy Society (ISES) have undertaken product research that has helped the renewable energy industry to grow. ISES, through its knowledge-sharing and community-building programs, helps its global membership provide the technical answers to accelerate the transformation to 100% renewable energy. ISES envisions a world with 100% renewable energy for everyone, with resources used wisely and efficiently. Founded in 1954. Location: Freiburg, Germany

Seas at Risk

www.seas-at-risk.org

Seas at Risk is one of the main environmental organizations in Europe tirelessly working to see a clean, non-polluted marine environment. Seas at Risk mainly focuses on the North and Irish Sea waters, as well as those of the larger North-East Atlantic. Location: Brussels, Belgium

Urban Climate Change Research Network — European Hub

www.uccrn-europe.org

The Urban Climate Change Research Network (UCCRN) European Hub promotes integrated climate change responses in cities based on knowledge-sharing and collaboration among European scholars, institutions, local governments, and industry while acknowledging the richness and diversity of Europe's scientists and practitioners. Through its interactive Urban Design Climate Workshops (UDCWs), the UCCRN European Hub sets the research and coordination agenda for the highest-priority climate change issues that European cities face. Founded in 2015. Location: Paris, France.

3E. AUSTRALIA AND OCEANIA

350 Pacific

www.350pacific.org

The organization 350 Pacific is a youth-led grassroots network working with communities to fight climate change in the Pacific Islands. It has facilitated workshops, organized days of action to raise awareness about climate change, participated in the UN climate negotiations, and is currently mobilizing the warriors of the Pacific Islands to challenge the fossil fuel industry. Location: Apia, Samoa.

Antarctic Ocean Alliance

www.antarcticocean.org

The Antarctic Ocean Alliance is a coalition of environmental and conservation organizations working to support the creation of a network of marine protected areas in Antarctica. They advocate for the Southern Ocean, the world's healthiest ocean, and the thriving populations of penguins, seals, and whales that live there. Founded in 2010. Location: Sydney, Australia.

Australian Association for Environmental Education

www.aaee.org.au

As Australia's leading professional body for environmental educators, the Australian Association for Environmental Education (AAEE) advocates for environmental education and assists with skills development to help to enable its members to excel in education for sustainability. The vision is to advance understandings and actions in relation to environmental and sustainability issues across all communities and education sectors. Founded in 1979. Location: Girrawheen, Western Australia.

Australian Conservation Foundation

www.acfonline.org.au

The Australian Conservation Foundation is Australia's national environmental organization. They are 700,000 people who speak out for the air we breathe, the water we drink, and the places and wildlife we love. This foundation is proudly independent, non-partisan, and funded by donations from the community. Founded in 1965. Location: Carlton, Victoria.

Australian Environment Business Network

www.aebn.com

The Australian Environment Business Network (AEBN) is Australia's leading organization that specializes in representing industry, business, and councils across Australia on environmental and carbon matters likely to impact any organization and business operations. Founded in 1999. Location: Altona, Victoria.

Australian Land Groundwater Association

www.landandgroundwater.com

The mission of the Australian Land Groundwater Association (ALGA) is to promote the protection, restoration, and management of land and groundwater for the benefit of human health and the broader environment across Australasia. Founded in 2007. Location: Robertson, New South Wales.

Australian Water Association

www.awa.asn.au

The Australian Water Association is Australia's largest water network that promotes sustainable water management. Founded in 1962. Location: St Leonards, New South Wales.

Banksia Foundation

www.banksiafdn.com

The Banksia Foundation is a not-for-profit organization focused on working with industry and the community to establish platforms that focus on the recognition of excellence in sustainability. Founded in 1989. Location: Melbourne, Victoria.

Ceres

www.ceres.org.au

Ceres acts with love and compassion in the face of the global climate and ecological emergency. Through powerful networks and advocacy, Ceres tackles the world's biggest sustainability challenges, including climate change, water scarcity and pollution, and inequitable workplaces. Founded in 1989. Location: Brunswick East, Victoria.

Certified Environmental Practitioner Program

www.cenvp.org

The Certified Environmental Practitioner (CEnvP) Program assesses environmental and social professionals in competency criteria of training, experience, professional conduct, and ethical behavior, and provides industry-wide certification and specializations. Founded in 2004. Location: Balywn, Victoria.

Climate Action Network Australia

www.cana.net.au

Climate Action Network Australia is a network that supports members and their allies to take actions to protect people at home and abroad from climate change, to safeguard our natural environment, and to build a fair, clean, healthy Australia for everyone. The vision — a fair and sustainable Australia free of climate pollution, where people and nature are protected from dangerous climate change. Founded in 1989. Location: Ultimo, New South Wales.

Climate Council

www.climatecouncil.org.au

The Climate Council is Australia's leading climate change communications non-profit organization formed to provide independent, authoritative information on climate change and its solutions to the Australian public. It advocates reducing greenhouse gas emissions. Founding in 2013.
Location: Potts Point, New South Wales.

Conservation Council of Western Australia

www.ccwa.org.au

The Conservation Council of Western Australia (CCWA) is the state's foremost non-profit, non-government conservation organization. CCWA is a small team of mostly part-time staff and volunteers, with an Executive Committee. Together, they represent more than 100 environmental organizations across Western Australia, which have joined as Member Groups. Founded in 1967. Location: Perth, Australia.

Environment Victoria

www.environmentvictoria.org.au

Environment Victoria is an independent charity, funded by donations. They have grown into a community of 40 grassroots member groups and more than 100,000 individual supporters. Together, they are campaigning to solve the climate crisis and build a thriving, sustainable society that protects and values nature. Founded in 1969. Location: Carlton, Victoria.

Environment Institute of Australia & New Zealand

www.eianz.org

The Environment Institute is a professional association for environmental practitioners from across Australia and New Zealand. The Institute supports environmental practitioners and promotes independent and interdisciplinary discussion on environmental issues. The Institute also advocates environmental knowledge and awareness, advancing ethical and competent environmental practice. Founded in 1986. Location: Melbourne, Victoria.

Gould League

www.gould.org.au

The Gould League is an independent not-for-profit organization celebrating over one hundred years of environmental and sustainability education. They help teachers to reinforce the impact of their science and sustainability curriculum, connect students with their natural world, and empower the community with positive messages and practical actions to live more sustainably. Founded in 1909. Location: Cheltenham, Victoria.

Port Phillip Emergency Climate Action Network

www.pecan.org.au

Port Phillip Emergency Climate Action Network (PECAN) is an alliance of 11 community groups who are working together on climate issues affecting this area. Founded in 2019. Location: Port Phillip, Victoria.

School Strike 4 Climate

www.schoolstrike4climate.com

School Strike 4 Climate Australia was started by three teenagers from Castlemaine, inspired by Greta Thunberg's weekly strikes outside Swedish Parliament. Since then, it has grown into one of the biggest movements in Australian history. School students of all ages, races, genders, backgrounds, and sexualities from every part of Australia are united by their concern for the future of the planet. They are striking from school to demand that politicians take the future seriously and treat climate change as what it is — a crisis. Founded in 2018.

Sustainability Victoria

www.sustainability.vic.gov.au

This organization supports Victorians to transition to a circular, climate-resilient economy. Sustainability Victoria is a Victorian government agency that delivers programs on integrated waste management and resource efficiency to the state of Victoria, Australia. It was established under the Sustainability Victoria Act 2005 to promote and facilitate environmental sustainability in the use of resources. Founded in 2005. Location: Victoria, Australia.

3F. ASIA

Asia Oceania Geosciences Society

www.asiaoceania.org/society/index.asp

Asia Oceania Geosciences Society (AOGS) works to promote geophysical science throughout Asia and Oceania. This international society publishes journals and has developed close connections with local and regional societies. Founded in 2003. Location: Singapore.

Asian Association on Remote Sensing

www.a-a-r-s.org

Asian Association on Remote Sensing (AARS) is an Asian non-governmental organization (NGO) dedicated to promote Remote Sensing through exchange of information, mutual cooperation, and international understanding and goodwill amongst the member countries of the Asia-Pacific region. Founded in 1981. Location: Tokyo, Japan.

Asian Disaster Reduction Center (ADRC)

www.adrc.asia

The mission of the Asian Disaster Reduction Center (ADRC) is to enhance the disaster resilience of its member countries, building safe communities, and to create a society where sustainable development is attainable. ADRC works to build disaster-resilient communities and to establish networks among countries through personnel exchanges and other programs. Founded in 1998. Location of Secretariat: Kobe, Japan.

Asia Pacific Network

www.apn-gcr.org

The Asia Pacific Network (APN) is an intergovernmental network of 22 countries working toward an Asia-Pacific region that is successfully addressing the challenges of global change and sustainability. Founded in 1996. Location of Secretariat: Kobe, Japan.

Cambodian Research Development Institute

www.cdri.org.kh

The Cambodian Research Development Institute (CDRI) is an independent Cambodian development policy research institute that works to contribute to Cambodia's sustainable development and the wellbeing of its people through the generation of high-quality policy-relevant development research, knowledge dissemination, and capacity building. Founded in 1990. Location: Cambodia.

Center for Environmental and Geographic Information Services

www.cegisbd.com

Initially created by the Bangladesh Ministry of Water Resources, the Center for Environmental and Geographic Information Services is a scientifically independent center of excellence. It uses integrated environmental analysis, geographic information systems, remote sensing, and information technology to manage natural resources for sustainable socio-economic development. Founded in 2002. Location: Dhaka, Bangladesh.

Clean Air Initiative for Asian Cities

https://www.cleanairinitiative.org/

The Clean Air Initiative for Asian Cities (CAI Asia) promotes and demonstrates innovative ways to improve the air quality of Asian cities through partnerships and information sharing. CAI Asia is a non-binding, multi-stakeholder network of government agencies, non-governmental organizations (NGOs), research institutes, international organizations, and private sector firms. Founded in 2001. Location: Pasig City, Philippines.

Geo-Informatics and Space Technology Development Agency

www.theknowledgeworld.com

The Geo-Informatics and Space Technology Development Agency (GISTDA) develops applications and provides data services relating to space technology and geo-informatics. It serves as the lead organization in Thailand for establishing common standards for remote sensing and geo-informatics systems. The organization has also created a natural resource information center. Founded in 2000. Location: Bangkok, Thailand.

Indian Council of Agricultural Research

https://icar.org.in/

The Indian Council of Agricultural Research (ICAR) serves as a repository of information and provides consultancy on agriculture, horticulture, resource management, animal sciences, agricultural engineering, fisheries, agricultural extension, agricultural education, home science and agricultural communication. It also coordinates agricultural research and development programs and to develop linkages at the national and international levels with related organizations to enhance the quality of life of the farming community. Founded in 1930. Location: New Delhi, India.

Indonesian Forum for Environment

https://www.walhi.or.id/

The Indonesian Forum for Environment (WAHLI) is the largest forum of non-government and community-based organizations in Indonesia, and works to promote social transformation, peoples' sovereignty, and sustainability of life and livelihoods. Founded in 1980. Location: Jakarta, Indonesia.

Institute for Global Environmental Strategies

https://www.iges.or.jp/en

The Institute for Global Environmental Strategies (IGES) is a strategic policy research institute that deals with global environmental issues. IGES research focuses on sustainable development in the Asia-Pacific, a region experiencing rapid growth in industrial activity and population, with serious implications for the future global environment. Founded in 1998. Location: Kanagawa, Japan.

International Rice Research Institute

www.irri.org

The International Rice Research Institute (IRRI) is a non-profit research and training center established to reduce poverty and hunger, improve the health of rice farmers and consumers, and ensure environmental sustainability through collaborative research, partnerships, and strengthening of national agricultural research and extension systems. Founded in 1960. Location: Los Banos, Philippines.

Kenan Institute Asia

www.kenan-asia.org

The Kenan Institute works to build sustainable competitiveness in Thailand and the Greater Mekong Sub-region. The Institute accomplishes this mission by developing multi-sector partnerships based on development needs and mutual benefit between Asia and the U.S. through the cooperation of universities, government agencies, and the private sector. Founded in 1996. Location: Thailand.

Malaysian Agricultural Research and Development Institute

https://landportal.org/organization/malaysian-agricultural-research-and-development-institute

The Malaysian Agricultural Research and Development Institute (MARDI) implements contract research and development (R&D) projects, and provides related technical and entrepreneurial development services relevant to food, agriculture, and related service industries. MARDI's technical services are in the form of advisory, consultancy, technical training, laboratory and quality assurance services, product and process development, and technology up-scaling. Founded in 1969. Location: Selangor, Malaysia.

National Science and Technology Development Agency

www.nstda.or.th/en

The National Science and Technology Development Agency (NSTDA) promotes research and development to strengthen Thailand's sustainable competitiveness, complemented with technology transfer and the development of human resources, as well as the development of scientific and technological infrastructure. Founded in 1991. Location: Pathum Thani, Thailand.

Network of Aquaculture Centres in Asia Pacific

www.enaca.org

The Network of Aquaculture Centers in Asia Pacific (NACA) is an intergovernmental organization that promotes rural development through sustainable aquaculture. The NACA seeks to improve rural income, increase food production and foreign exchange earnings, and to diversify farm production. The ultimate beneficiaries of NACA activities are farmers and rural communities. Founded in 1988. Location: Bangkok, Thailand.

Sri Lanka Council for Agricultural Research Policy
www.slcarp.lk

The Council for Agricultural Research Policy (CARP)'s main function is to develop a vibrant, effective and sustainable system of agricultural research, promoting socio-economic development in Sri Lanka. They work to ensure agricultural research, development, and innovations are directed towards national development goals through policy formulation, facilitation, coordination, monitoring and evaluation, and impact assessment. Founded in 1987. Location: Sri Lanka.

Stockholm Environment Institute–Asia Centre
www.sei.org/centres/asia/about

SEI-Asia has a diverse team of multinational experts that integrates scientific research with participatory approaches to co-develop and share knowledge, build partnerships, and influence policy for resilient development. It focuses on gender and social equity, climate adaptation, disaster risk reduction, water insecurity and integrated water resources management, transitional agriculture, renewable energy, and urbanization. Founded in 2004. Location: Bangkok, Thailand.

Tata Energy and Resources Institute
www.teriin.org

TERI's mission is to usher transitions to a cleaner and sustainable future through the conservation and efficient use of energy and other resources, and innovative ways of minimizing and reusing waste. Founded in 1981. Location: New Delhi, India.

The Institute of Geographical Studies
www.tigs.in

The Institute of Geographical Studies (TIGS) believes that geography is a vital tool in developing skills which will help people become activism-oriented.

TIGS works with a wide range of audiences including students, teachers, researchers, organizations, and civil societies. Founded in 2015. Location: Bangalore, India.

3G. AFRICA

BirdLife South Africa

www.birdlife.org.za

BirdLife South Africa strives to conserve birds, their habitats, and biodiversity through scientifically based programs that support the sustainable and equitable use of natural resources and encourage people to enjoy and value nature. Founded in 1996. Location: Johannesburg, South Africa.

Botanical Society of South Africa

www.botanicalsociety.org.za

The Botanical Society of South Africa's mission is to win the hearts, minds, and material support of individuals and organizations, wherever they may be, for the conservation, cultivation, study, and wise use of the indigenous flora and vegetation of South Africa. Founded in 1913. Location: Cape Town, South Africa.

Centre for Environmental Rights

www.cer.org.za

At the Centre for Environmental Rights, activist lawyers work with communities and civil society organizations in South Africa to realize the Constitutional right to a healthy environment by advocating and litigating for environmental justice. The group seeks a just, equitable, compassionate society that is resilient, celebrates diversity, and respects the interdependence between people and the environment. Founded in 2009. Location: Cape Town, South Africa.

Endangered Wildlife Trust

www.ewt.org.za

The Endangered Wildlife Trust works to help safeguard threatened species and ecosystems in southern Africa with the vision of helping humans and wildlife prosper in harmony. Its projects target helping all sorts of endangered species from vultures to rhinos, with various conservation programs at targeted habitats. Founded in 1979. Location: Gauteng, South Africa.

Wilderness Foundation Africa

www.wildernessfoundation.co.za

Wilderness Foundation Africa (WFA) works to protect and sustain wildlife and wilderness through integrated conservation and education programs. Whether it is direct action anti-poaching in the field, large landscape wilderness management, or developing rising young leaders from disadvantaged communities for a career in conservation. Founded in 1973. Location: Port Elizabeth, South Africa.

Wildlands Conservation Trust

www.wildtrust.co.za

The Wildlands Conservation Trust (Wildtrust) is a non-profit organization that works to conserve the natural heritage of South Africa. Wildtrust acts to ensure the safety of threatened species, conserving and restoring the ecosystems in which they can thrive, and the upliftment of people and communities. Founded in 2004. Location: Pietermaritzburg, South Africa.

Wildlife and Environment Society of South Africa

www.wessa.org.za

The Wildlife and Environment Society of South Africa (WESSA) is a South African environmental organization which aims to initiate and support high impact environmental and conservation projects to promote participation in caring for the Earth. For over 90 years, WESSA has proactively engaged with the challenges and opportunities presented by South Africa's unique natural heritage and the social and economic systems that depend on it. Founded in 1926. Location: Howick, South Africa.

World Wildlife Fund for Nature South Africa

www.wwf.org.za

From the impact on the fast-changing climate to the free-flowing rivers, expansive life-giving oceans to vast food supplying landscapes, the World Wildlife Fund for Nature (WWF South Africa) catalyzes strategic initiatives where there is the greatest need to restore balance, reduce impact, and protect the country's vital resources and natural biodiversity. Founded in 1968. Location: Cape Town, South Africa.

3H. LATIN AMERICA

Ashoka Argentina

www.ashoka.org/es-ar

Ashoka champions and provides scholarships to activists who are committed to social change. A global organization, Ashoka works with social entrepreneurs whose projects have addressed innovative solutions to problems in education, health, environment, diversity, and citizen participation. Founded in 1994. Location: Buenos Aires, Argentina.

Avina Foundation

https://www.avina.net/en/home/

Rooted in the global South, Fundación Avina works to drive collaborative processes that bring about systemic changes in favor of human dignity and care for the planet. Founded in 1994. Location: Panama City, Panama

Charles Darwin Foundation

www.darwinfoundation.org/en

The Charles Darwin Foundation for the Galapagos Islands (CDF) was founded by a team of worldwide conservationists to provide scientific solutions that protect the Galapagos Islands. The non-governmental organization has protected many species on the islands from going extinct, helped set up marine preserves, and trained many Galapagueño and Ecuadorian students. Founded in 1959. Location: Galapagos, Ecuador.

Comité Nacional Pro-Defensa de la Flora y Fauna

www.codeff.cl

The Comité Nacional Pro-Defensa de la Flora y Fauna (CODEFF) Chile works to inspire environmental responsibility in the Chilean society in order to protect the country's natural resources. Their programs range from protecting Humboldt Penguin colonies to creating marine sanctuaries and managing protected areas. Founded in 1968, the non-governmental organization plays a central role in safeguarding Chile's environmental heritage. Founded in 1968. Location: Santiago, Chile.

Confederación de Nacionalidades Indígenas del Ecuador

www.conaie.org

The Confederation of Indigenous Nationalities of Ecuador (CONAIE) is a non-governmental organization that brings together indigenous communities in Ecuador in order to protect their land rights, culture, and language. Collectively, the organization seeks equality and justice for indigenous peoples and channels their power into civic participation in order to create lasting social change. Location: Quito, Ecuador.

Conservação Internacional Brazil

www.conservation.org/brasil

Conservation International is a Brazilian non-governmental organization that works to ensure a healthy and productive planet for all. A part of a global network, in Brazil this NGO focuses on protecting regions such as Três Fronteiras, Matopiba, and the Paraguaçu Basin, and on the issues of sustainable agriculture, indigenous people's rights, public policy, and climate change. Founded in 1990. Location: Rio de Janeiro, Brazil.

Conservación Patagónica

www.conservacionpatagonica.org/#modal

Merged with the Tompkins Conservation, Conservación Patagónica works to create national parks in Patagonia in order to protect wildlands and wildlife and generate healthy economic opportunities for local communities. With just 5% of Patagonia under protection (compared to a global average of 14%), the region needs formal conservation and safeguarding against unchecked corporate development. Founded in 2000. Location: Valle Chacabuco, Chile.

Greenpeace Brazil

www.greenpeace.org/brasil/

Greenpeace established a presence in Brazil in the early 1990s, and its first victory was helping create a ban on the import of toxic waste. Since then, the non-governmental organization has been campaigning diligently against illegal and predatory logging in the Amazon, increasing awareness about reducing greenhouse gas emissions, and pressuring the government to encourage alternative energy. Founded in 1992. Location: São Paulo, Brazil.

Iwokrama International Centre for Rainforest Conservation

www.iwokrama.org

Located in the heart of the Iwokrama rainforest, the Iwokrama International Centre for Rainforest Conservation advocates selective timber harvesting, eco-tourism, and conservation research. The non-governmental organization also works to help conserve the North Rupununi, a natural area in southern Guyana that, for the last 30 years, has been isolated from the public eye. Founded in 1996. Location: Georgetown, Guyana.

United Nations Population Fund Pacific Agency Bolivia

www.bolivia.unfpa.org

The United Nations Population Fund Pacific Agency (UNFPA) works so that every pregnancy is wanted, every birth is safe, and every young person's potential is fulfilled. The non-governmental organization focus on sexual and reproductive health and rights, prevention of sexually transmitted infections, gender equality, and emerging issues in population and development, such as ageing and climate change. Location: La Paz, Bolivia.

Vida Silvestre Uruguay

www.vidasilvestre.org.uy

Vida Silvestre Uruguay is a conservation organization that is building a network of voluntary conservation initiatives on private farms (*refugios*) in order to protect Uruguay's biodiversity. The non-governmental organization also offers environmental education classes for young people and provides resources to civil society on the subjects of ecotourism and environmental legislation. Founded in 1995. Location: Montevideo, Uruguay.

Voluntarios Sin Fronteras

www.voluntariossf.org.ar

Founded by university students, Voluntarios Sin Fronteras promotes volunteerism and makes it easy for non-governmental organizations to connect with potential volunteers. The NGO promotes both domestic and international volunteer programs, and places a special focus on recruiting volunteers for schools and food banks, computer training, and sustainable development programs. Founded in 2005. Location: Buenos Aires, Argentina.

World Wildlife Fund Perú

www.wwf.org.pe

The mission of World Wildlife Fund (WWF) Perú is to conserve nature and reduce the most pressing threats to the diversity of life on Earth. With a global focus on climate change and the conservation of wildlife and wildlands, WWF Perú also works to protect the Amazon and its inhabitants and to educate Peruvian society on how to reduce their ecological footprint. Founded in 1998. Location: Lima, Perú.

Acknowledgments

I thank my three sisters — Alice Turnbull, Martha Bustin, and Cynthia Rosenzweig — who each contributed to the conception, evolution, and realization of this book.

Abby Luby, Chief Editor at ClimateYou.org and Director of Educational Programs at the ClimateYou Alliance, Inc., plays a major role in the development and operation of the ClimateYou.org website, and thus the foundation of this book. Her research formed the basis for the U.S. Presidents and Conference of the Parties timeline tables. It has been a pleasure to work with her on the website and this book.

Lisa Fine helped to create the resource list of climate change organizations around the world. I appreciate her work on this extensive list, which highlights the global community that is mobilizing to respond to the climate crisis.

Professor Reginald Blake of New York City College of Technology (City Tech), part of the City University of New York (CUNY), has shown expert leadership in the development of the ClimateYou Alliance City Tech Education Program, bringing the voices of more than 650 students forward on climate change. I thank Dan Bader and Somayya Ali Ibrahim for their early contributions to the ClimateYou.org website and the ClimateYou Alliance City Tech Education Program.

Eric Schultz provided key information about the intricacies of climate change and air conditioning. David Rind of NASA GISS did a helpful review of the James Hansen blog. David Soll gave near-real time insights on the climate bill, just as the book was going to press. Thank you to all three.

Juan Esquerra and Juliana Velez Duque did a fantastic job on the design and development of the new ClimateYou.org website. Juliana designed the striking ClimateYou.org banner and logo. Hannah Rosenzweig expertly and kindly prepared the author photo for publication.

At World Scientific Publishing, I am very grateful to Senior Executive Publisher Zvi Ruder, Senior Editor Amanda Yun, and the great team who produced the book.

Finally, I thank Andrew Bustin for his stunning photographs that so beautifully illustrate the five dimensions of the climate crisis.

Index

About the Author

After graduating from Dartmouth College, George Ropes joined the Peace Corps, serving in the Dominican Republic. He then taught in public and private schools in New York City before leaving to pursue his own advanced studies. At the Massachusetts Institute of Technology, he earned a master's degree in political science, with a concentration in International Nutrition Policy and Planning. He joined Catholic Relief Services (CRS), a non-profit aid and development organization. As a CRS program leader and specialist in the areas of food distribution and computerization, he lived and worked in Egypt, India, and Angola. In 2008, he co-founded the website ClimateYou.org, a collaborative forum designed to connect climate researchers, writers, students, and the general public. The site's initial and continuing goal: to help people share news and analysis of developments as the world wakes up to climate change and moves forward to solve its multidimensional challenges. George Ropes lives in Wilmington, Delaware.